FRANCO RAGNI

GERMAN FIGHTERS
of WORLD WAR II

squadron/signal publications

PUBLISHED 1979 by SQUADRON/SIGNAL PUBLICA-
TIONS, INC.
1115 CROWLEY DRIVE, CARROLLTON, TEXAS 75006

ISBN 0-89747 - 105-9

Printed in Italy
La Tipografica Parmense

Designed by:
Corrado Barbieri

Color silhouettes and line drawings by:
Pietro Mazzardi
Silvio Lora-Lamia
Franco Ragni
Richard Caruana
Marco Gueli
S. Jamois
Gianni Riccardi
Enzo Maio

Cutaway drawings by:
Roberto Marvasi
Alfonso Rigato

LIST OF CONTENTS

Me 109G-6 at take-off.

INTRODUCTION

When the Treaty of Versailles was signed in June 1919, its Air Clauses seemed so severe as to permanently exclude any possibility of a resurrection of German military aviation. The clauses were indeed severe, ordering the complete neutralization of all German military aircraft and prohibiting all future aircraft production. An Inter-allied commission set up for that purpose, was to ensure, as indeed it did, that the terms of surrender were scrupulously observed. The construction of civil aircraft was not, however, banned, although for some years the size of such craft was limited.

Furthermore, the various German industries intending to continue aircraft production created subsidiaries abroad (where they were free from any restriction). Lastly, it was absurd to think that restrictions of the kind would be maintained for long against an industrial power such as Germany; and in fact, after a few years, for both political and economic reasons, the Control Commission was already behaving in a more tolerant manner.

All these factors meant that German aeronautical technology was able to keep up with the progress of those years, thus maintaining the 'avant-garde' position it had held at the end of the First World War. In Germany, in fact, the techniques of constructing aircraft entirely in metal (including the skinning) and of cantilever wing structure were already widespread. After the War this characteristic of technical innovation was seen above all in the designs produced by the Junkers and Dornier companies.

The other big problem, that of training personnel, was solved by collaborating with the Soviet Union, the other great nation which found itself in a state of political and economic isolation in relation to the Western Powers. It was inevitable that the two 'great frustrated powers' should unite and study plans for a common program. The Soviet Union needed to rebuild its own, armed forces, both in men and materiel and, in the secret clauses of the Treaty of Rapallo (drawn up by the two countries in 1922), the Germans agreed to help them in return for training facilities for their own men. They participated, furthermore, in the construction and research testing of aircraft for the Soviet Air Force, aircraft which were often of German conception.

Later, around the mid-30s when the Nazis came to power, the similarity of their ideology to that of the Fascist régime in Italy, as well as the international prestige which Italian aviation enjoyed during that period, led to the clandestine training of many German pilots in Italy, too. At the same time, the Germans withdrew from the Soviet schools, the situation now being advanced enough for the German Luftwaffe (Air Force) to cease existing in secret; the Russians, too had now become 'of age' and the German presence was becoming awkward.

Hitler became Chancellor of the Reich in the Spring of 1933 and entrusted the position of 'Shadow-Minister' for Aviation to Hermann Göring. The Luftwaffe officially came into being on 1 March 1935 with Hitler declaring to the world on the occasion that the considered himself free of any obligation regarding armaments. Immediately, the as-yet-young Luftwaffe began an intense propaganda campaign, seeking to give the impression of a numerically much greater force than it had in reality. The first fighter Geschwaders of the new Luftwaffe were equipped with Heinkel He 51 biplanes soon supplemented by Arado Ar 68s. They were adequate for the time and had an acceptable performance record, even if the He 51 during the Spanish Civil War (as part of the German Condor Legion) didn't demonstrate exceptional qualities, especially when compared to the Italian CR32 and the Soviet I-15. Both the German fighters had 12-cylinder in-line engines, and this was to be a permanent feature of German technology, with few exceptions. German engine production was to settle on two series of liquid cooled inverted V engines made by Junkers (Jumo) and Daimler-Benz, and possessing an excellent weight power ratio, undoubtedly superior to that of its contemporary, the Hispano Suiza 12 Y, but not achieving the outstanding quality of the Rolls Royce Merlin. The possession of such engines, plus the natural German tendency for making technically advanced choices, and their formidable organizational capacities, meant that in the space of a few years they had created a first-class air force from both the quantitative and qualitative stand point. A very good example, still in the area of fighter planes, was the production of the Messerschmitt Bf 109 and Bf 110. There were, however, some negative aspects, such as the excessive importance given to personal and political sympathies and jealousies in the relations between the different authorities which controlled decision-making at various levels. This resulted in choices which often followed rather incoherent courses. The natural organizing character of the Germans led them to carry out diverse and frequently contradictory manufacturing instructions, all following each other within a brief space of time, with the same determination and, obviously, a great waste of energy and resources.

These factors became decisive in the later phases of the War, when it was even more necessary to rationalize efforts, and culminated in an absurd proliferation of military and Party bodies, all competing against each other for control of the organization of the Reich's defence forces. Under these conditions, the fighter corps was the one to feel its most negative effects.

Another negative factor which damaged the German fighter force was the hesitation in adopting a defensive mentality in the first part of the War and later (as late as 1944!) the continued insistence on giving absolute priority to the manufacture of bombers. If the first was understandable (in the early years the German forces had always been victorious on all fronts), the second was to prove fatal.

In spite of everything, the German fighter force enjoyed notable operational prestige, was among the bestequipped forces existing in the world and obtained really exceptional results. The planes were always abreast of the times even in the worst cases and the organisation was such that even the smallest command units had an autonomy which permitted a high numerical efficiency in relation to the importance of the division. This explains how relatively small expeditionary forces (such as those sent to the Mediterranean front) were able to maintain such a high level of service. The capacity of the Germans to adapt their planes (and themselves) to the most diverse and unpredictable operational situations was also notable. It is sufficient to remember how quickly they brought out tropical versions of their planes for use in Africa, whilst the Italians, who had had planes in Africa since 1911, were still sending whole fighter plane divisions to Libya without dust filters.

The Germans also 'invented' a new category of fighter plane (even though understandably a little rough and ready at the beginning): the twin-engined heavy fighter.

The first modern version of a similar type was in fact the Messerschmitt Bf 110, born in 1936 to be one of the special 'zerstörer' (destroyer) aircraft which Goering predicted would be the decisive weapon in the war of the air. In fact, the new aircraft failed in some of the tasks envisaged for it, such as that of a strategic fighter for escorting the bombers which flew deep into enemy territory, (the Battle of Britain being the most resounding failure). But the Bf 110 (and all its successors for whom it served as a basis) remained one of the work horses of the German Fighter Force, leaving a deep mark on the history of air warfare with its adaptability to the most diverse rôles.

It was fortunate for the Luftwaffe that the pre-war designed fighters were so advanced technically, as they remained fundamentally the same type up to the last battles in 1945, with the exception of a few jet fighters which were hurriedly brought into service in the last months of the War.

In fact, during the whole conflict they did not succeed in bringing any really new type of fighter to a satisfactory operational efficiency level (the same can be said of the bombers, too), probably owing to excessively demanding ministerial specifications that even the most inventive of the German designers were not able to satisfy.

In concluding this introduction to Second World War German fighter aircraft, something must be said about their pilots, another exceptional 'product made in Germany'. Even if in the last moments of the War the average level of quality of German fighter pilots dropped because of insufficient training given to the new pilots, the global results of the War were probably unequalled. This can be seen by simply looking at the list of 'aces' and their victories: 103 pilots with more than 100 victories, and of these, 13 with more than 200 and, just think of it, 3 with more than 300 victories! An exaggeration? No more than with the other air forces; the system of confirmation was very strict and the exceptional number of victories can be explained quite satisfactorily when one considers the continual high quality of the planes and their weapons (as opposed to an enemy often less well-equipped) the experience and aggressive nature of the pilots, and their very long, tireless service, strangers to rest periods and other pleasures of that kind. To summarize briefly the exceptional results of the Luftwaffe fighter pilots, let us express the numerical terms, reported above, in another way (taking into account over-estimates of the claims): the 103 pilots quoted, alone accumulated a total of at least 10-15,000 enemy aircraft destroyed! The great enemy aircraft destruction was at its highest on the Russian front especially in the first years of the German offensive there, but even on the Western Front exceptional victories weren't lacking: take for example the 103 victories of Adolf Galland, all gained before December 1941, when he was nominated General of the 'Jagdflieger' (hunter pilots).

The ace of aces was Major Erich Hartmann, with

352 victories (7 of which were on the Western front and 345 on the Russian) results indirectly confirmed by their similarity to figures found in the Sentence of the Soviet Military tribunal which condemned the Major to 20 years hard labor, for 'sabotage' of the Soviet economy by destroying 352 planes.

This introduction may seem excessively laudatory and in fact a thorough study has not yet been made. What we would like to underline is the fact that the German fighters always aroused great respect and fear in the enemy throughout the War, and even if the general panorama shows some 'shadows' among the many 'lights' probably no-one could have done better under the same conditions. Even today the Second World War Air Force of Germany, particularly the fighter force, constitutes a subject of great historical interest, and we intend to make our contribution in this area.

MESSERSCHMITT Bf 109 series

The Messerschmitt 109 is perhaps the most famous fighter plane of all times. The 'Messerschmitt' by name was admired and feared by fighter pilots of the Second World War, who abbreviated its name to 'Messy' (GB), 'Messer' (Fr) and 'One-O-Nine' (USA). Although the record is contested by the Soviet 'Shturmovik', its probable that, including the models made in Czechoslovakia and Spain, the Messerschmitt remained for many years the plane produced in the greatest numbers in history. But this is not the only reason why the Bf 109 merits a place of high standing in aviation history.

The family tree of the Bf 109 is quite well known. The Bayerische Flugzeugwerke (Bavarian Aircraft Company), of which Willy Messerschmitt was co-manager, was given the job in 1933 of making a project study for a light competition aircraft, intended for the German participation (and if possible victory) in the Challenge de Tourisme Internationale programmed for the following year. From the design (already in production) of a light training monoplane known as the M37, the Bf 108 was born, incorporating the best that modern technology could offer in the field of aeronautical design; this was to remain a model of how a modern touring plane should be, and would remain valid for many years. The Bf 108, later known as the 'Taifun'

(Typhoon) was used as the basic design for the Bf 109 fighter project, created from a specification issued by the Luftwaffenführungstab for an all-metal monoplane fighter. Its competitors were the Heinkel He 112, the Arado Ar 80 and the Focke Wulf Fw 159. The most advanced technically was the Messerschmitt design with its entirely metal monocoque structure with formed stressed skin with flush riveting, automatic Handley Page slats on the wing leading edge, trailing edge flaps, closed cockpit, cantilever wing and retractable under-carriage. These features were not in themselves a novelty but it was the first time ever that they had been found together in one single airplane. Furthermore, the whole design was so well co-ordinated and far-reaching that this plane was still well up-to-date 10 years later.

For the fighters participating in the competition, in-line engines under development at Junkers and Daimler Benz were to be used, but none of them were finished in time for installation in the Bf 109a prototype, later named the Bf 109 V1 (thus inaugurating the new regulation regarding the numbering of experimental prototypes with

At the top: Bf 109D-1 'Dora' of 1/JG137 at Bernburg in 1939. Above left: a Bf 109D-1 built under license by Focke-Wulf, photographed in 1938 leaving the factory at Bremen. On the right: a Bf 109B-1 photographed in flight, in 1937. Left: a propaganda photo in which can be seen an old Bf 109E-1 or C-1 with the manufacturer's color 'RLM Grau 02'.

the designation 'Versuchs'). The first flight took place in Augsburg in September 1935, with a Rolls Royce 'Kestrel' engine of 695 HP; flying it was the chief test-pilot of Messerschmitt, 'Bubi' Knoetsch.

The plane aroused a lot of attention with its ultramodern features, and was at first treated with great diffidence, but finally, it was declared the competition winner after various complex events which were more associated with political maneuvering and historical tradition than aeronautical technology.

The next two prototypes, the Bf 109 V2 and V3 were noteworthy only for their Jumo 210A engines of 610HP. The latter was also armed as per specifications with 2 MG 17s of 7.9mm. The firepower was increased by a third machine-gun

9

Top: a Bf 109E-4/N (Trop) of I/JG27 'Afrika'. Above: maintenance personnel at work on the engine of a Bf 109E-7. Below: a Bf 109E (Emil) in maintenance.

experimental planes was used in Spain for a brief spell, contributing to the collection of information for development of future fighters.

In fact, during the Spanish conflict, all the versions of the Bf 109, as they came into service, were poured into this precious training ground for war of the renewed Luftwaffe.

The first production models were the Bf 109B-1 with a fixed pitch wooden twin-blade propeller, and the Bf 109B-2 with a variable pitch twin-blade metal propeller. Both were powered by 600HP Jumo 210Da engines. From the Spring of 1937 they were delivered to the first Jagdgruppen and almost immediately were sent to Spain to replace the obsolescent Heinkel He 51 biplanes of the Condor Legion. The armament consisted of two 7.9 mm synchronized machine-guns.

The last model with the Jumo engine was the Bf 109C, as the subsequent types used the more promising Daimler-Benz engines. The Bf 109C used the 700HP direct fuel injection Jumo 210G engine, had R/T radio equipment and was more heavily armed, with two wing machine-guns as well as the two in the fuselage. The Bf 109C-2 was also equipped with a fifth MG 17, between the cylinder banks of the engine. The Bf 109C was sent to Spain in small quantities to complete the replacement of the He 51. Because of delays in delivery of the Daimler-Benz engines, production of the Bf 109C continued into 1939. Ten models were also ordered by Switzerland, the sole example of exportation of a Bf 109 with the Jumo engine, but these were hybrids using Jumo 210D engines of the Bf 109B.

firing through the propeller shaft, in the next model, the Bf 109 V4. This, the prototype for the B version, like the V5 and V6, all used various versions of the Jumo 210. This trio of

Left: refuelling of a Bf 109E-1 of the 4/JG54 'Gruenherz'.
Above: a demonstration of the effectiveness of camouflage
on this Bf 109E-4/N of I/JG27.

Above: a Bf 109E-1 photographed probably in 1940.
Below: a Bf109E-7 of 7/JG26 'Herzas'. From a recently
published color photo, it would seem that the engine
cowling was painted in 'Gelb 27' or, more probably, in
'Weiss 21'.

The international fame of this exceptional fighter of the Messerschmitt company, carefully fueled by the propaganda organs of the Nazi régime, began to grow in this period due to a series of sporting exploits; a group of prototypes with the Jumo engine was victorious in the International Flying Meeting in Zurich in the Summer of '37 when they massacred all opposition in the speed event. Their activities in Spain, on the other hand, were wrapped in a veil of secrecy, but the more attentive observers could not fail to notice the decisive presence of the German fighter in the Nationalist ranks, even though the Jumo version's performance was not so different from that of the

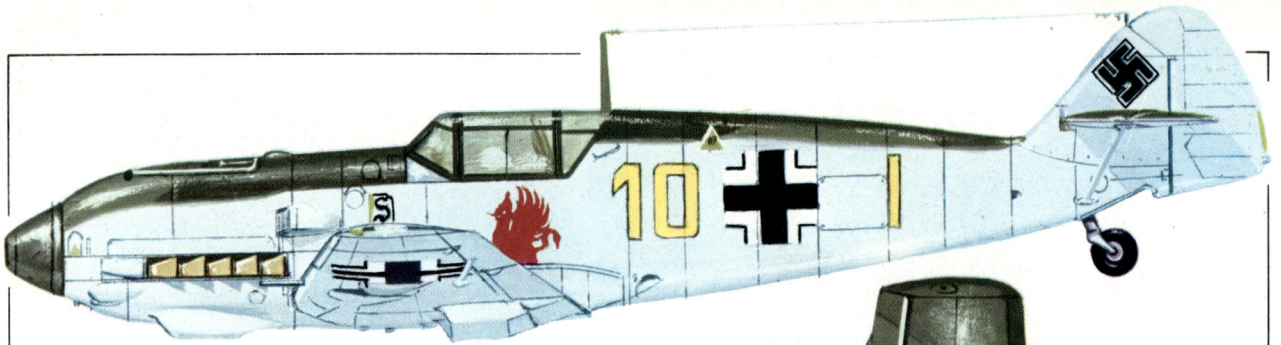

Messerschmitt Bf 109E 9/JG-26 'Hoel-
lenhund'-Caffiers, France - August
1940.

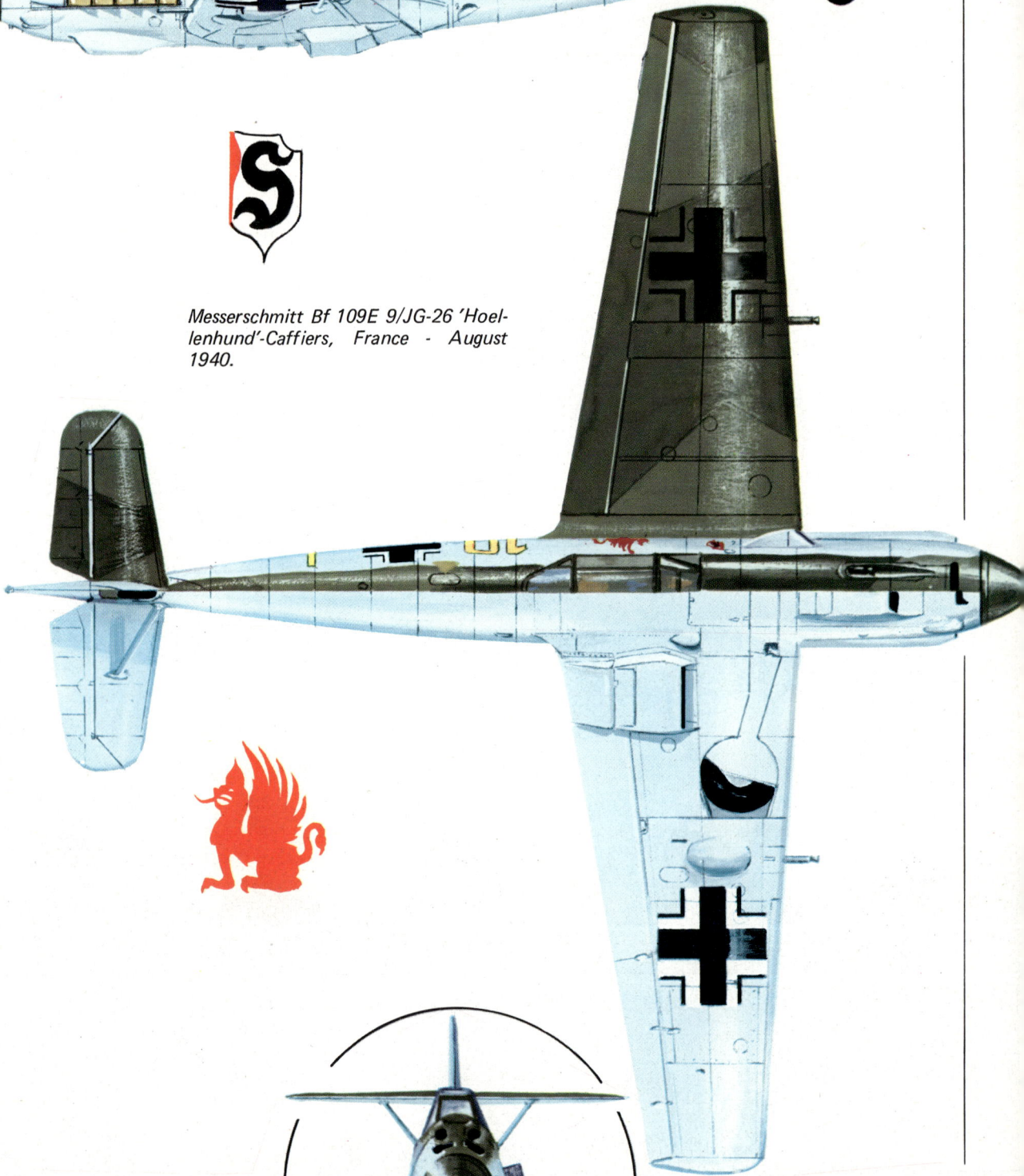

Richard J. Caruana

Bf 109E II/JG 54 - November 1939

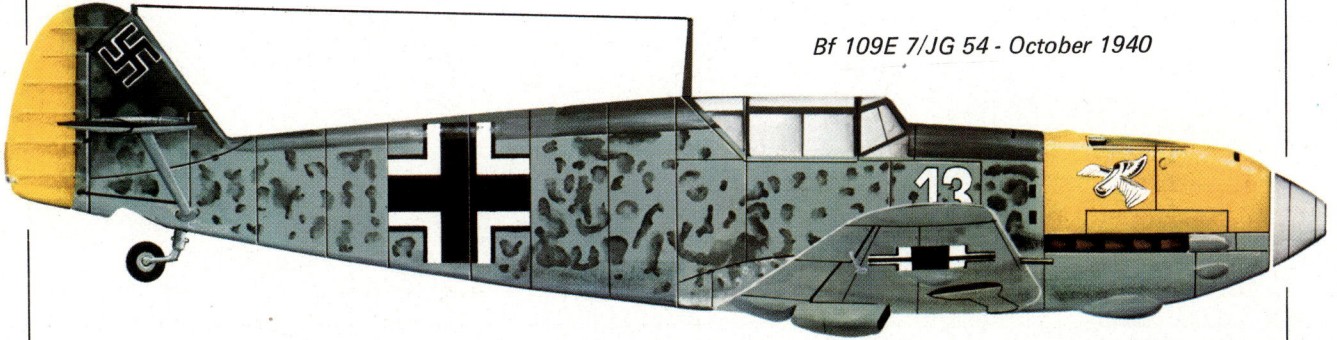

Bf 109E 7/JG 54 - October 1940

Bf 109E I/JG 52 - August 1940

Bf 109-4 (N) I/JG - Lybia 1941

Bf 109E-3 III/JG 27 - Balkans 1941

S. Jamois

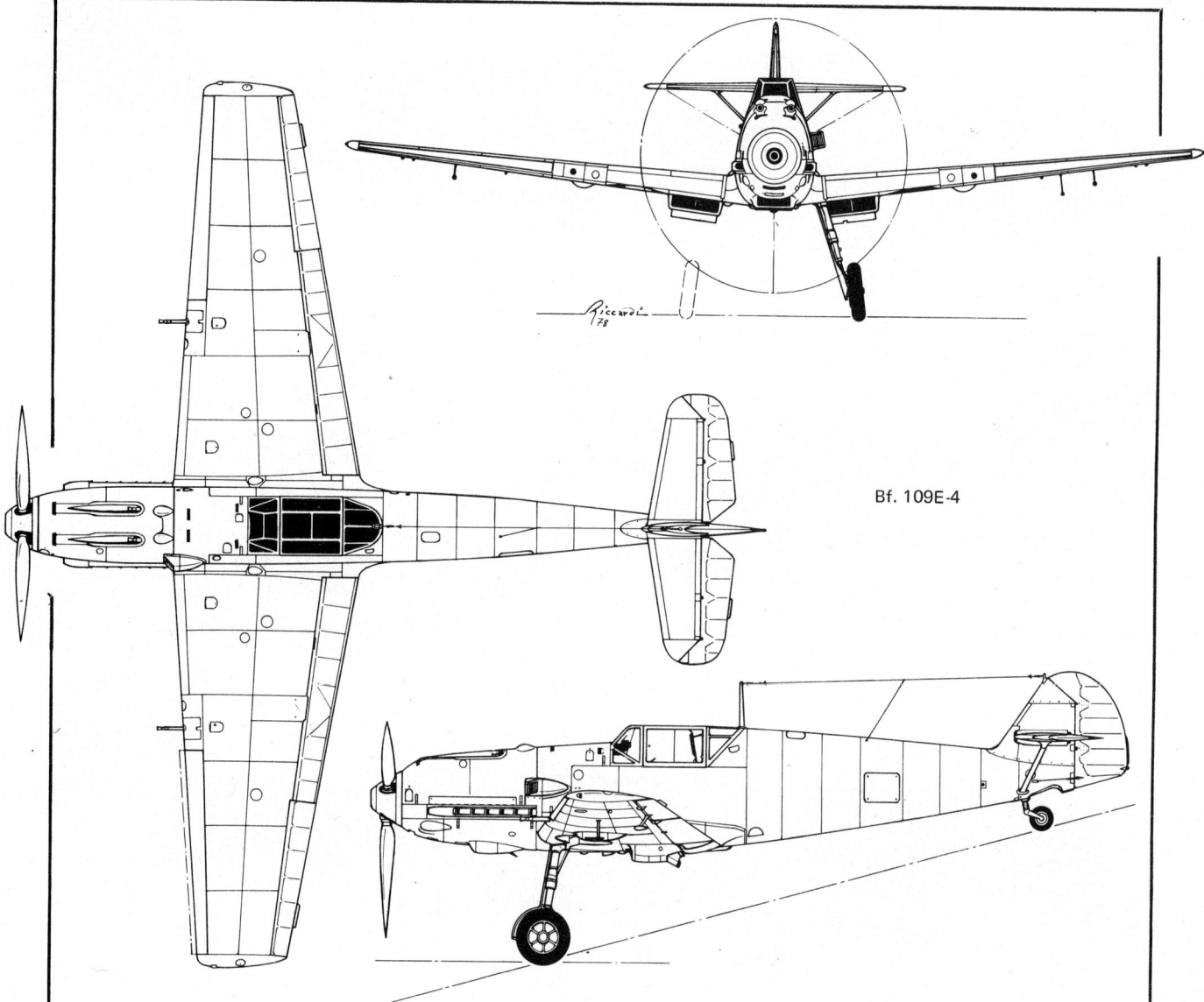

Bf. 109E-4

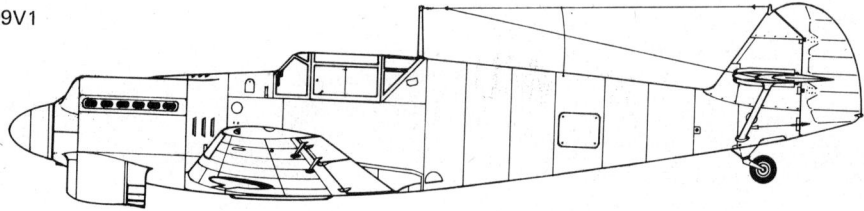

Bf. 109V1

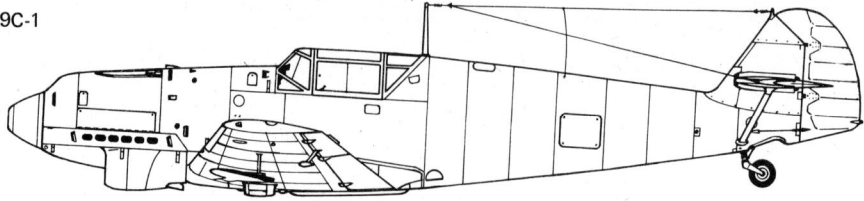

Bf. 109C-1

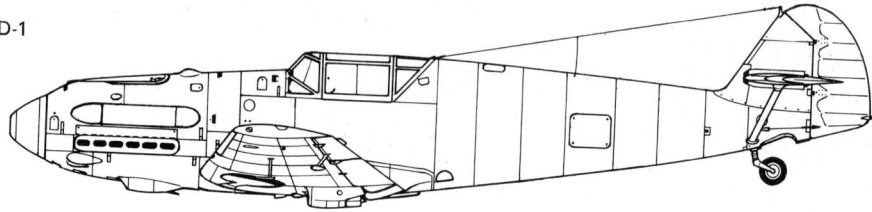

Bf.109D-1

At the top of the page: two Bf 109Es intercept a German multi-engined plane during an exercise. Above left: a Bf-109E with the early type of windshield and canopy. On the right: the 44th 'kill' is painted on a Bf 109F-1 on the Eastern front.

Soviet 'Rata' in service with the Republican airforce (the Bf 109B-2 had a maximum speed of 465 km/h-289mph).

The Bf 109D differed from its predecessors in its DB 600Aa engine of 986 HP using a new three-blade propeller of the same design as first tested on the Bf 109V10, originally with the Jumo engine. The Bf 109D was a transition version while awaiting the E version with the DB 601 direct fuel-injection engine more adapted to the needs of a fighter plane. The only mass-produced variant of the D series was the D-1, armed with the usual MG17 7.9 mm guns in the fuselage with, in addition, but not always, an MG/FF engine-mounted 20 mm cannon.

The first real mass-produced version was the Bf 109E, the crowning example of all the previous versions and an airborne symbol of the 'Golden Season' of the Luftwaffe. As has already been stated, the 'raison d'être' of this new versions was the availability of the 1050 HP DB 601A engine, an example of which had been tested on a D airframe, numbered V1, in 1938.

The previous year, the Bf 109V13 had been fitted with a specially boosted DB 601 to beat the speed record for landplanes, reaching 611 km/h. After a

Messerschmitt Me 109G-4/R3 (Gustav) cutaway drawing key

1 VDM three-blade propeller
2 MG151/20 cannon muzzle
3 Propeller hub and pitch changing gear
4 Oil tank
5 Daimler-Benz DB 605A-1 engine
6 MG17 machine gun muzzle
7 Wing leading edge
8 Pneumatic plumbing

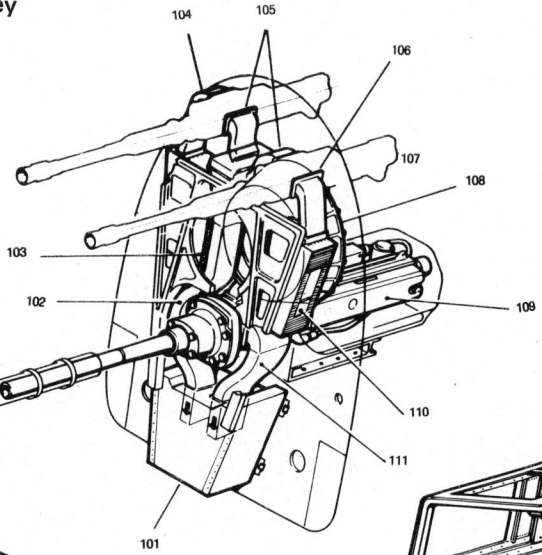

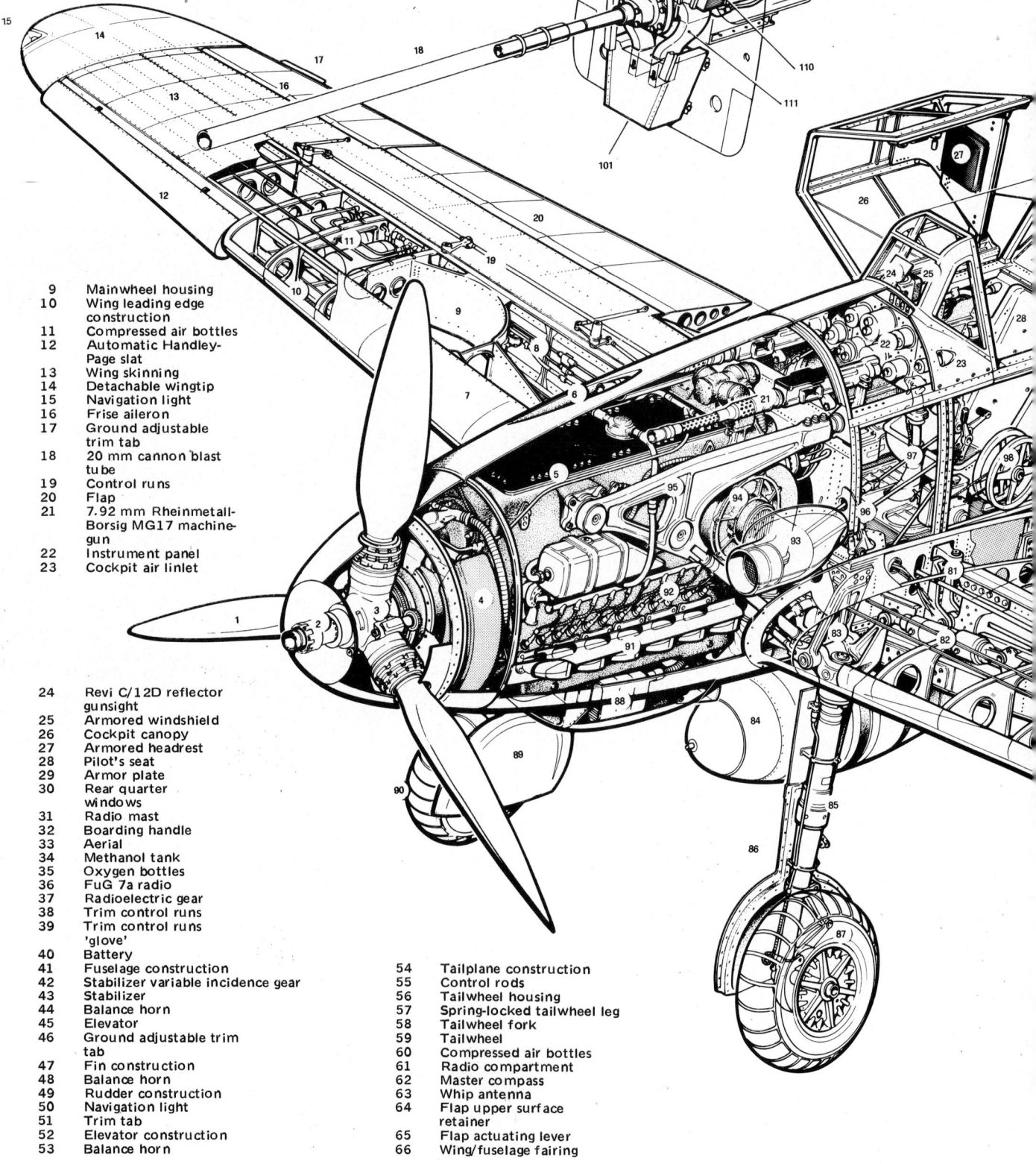

9 Mainwheel housing
10 Wing leading edge construction
11 Compressed air bottles
12 Automatic Handley-Page slat
13 Wing skinning
14 Detachable wingtip
15 Navigation light
16 Frise aileron
17 Ground adjustable trim tab
18 20 mm cannon blast tube
19 Control runs
20 Flap
21 7.92 mm Rheinmetall-Borsig MG17 machine-gun
22 Instrument panel
23 Cockpit air linlet

24 Revi C/12D reflector gunsight
25 Armored windshield
26 Cockpit canopy
27 Armored headrest
28 Pilot's seat
29 Armor plate
30 Rear quarter windows
31 Radio mast
32 Boarding handle
33 Aerial
34 Methanol tank
35 Oxygen bottles
36 FuG 7a radio
37 Radioelectric gear
38 Trim control runs
39 Trim control runs 'glove'
40 Battery
41 Fuselage construction
42 Stabilizer variable incidence gear
43 Stabilizer
44 Balance horn
45 Elevator
46 Ground adjustable trim tab
47 Fin construction
48 Balance horn
49 Rudder construction
50 Navigation light
51 Trim tab
52 Elevator construction
53 Balance horn

54 Tailplane construction
55 Control rods
56 Tailwheel housing
57 Spring-locked tailwheel leg
58 Tailwheel fork
59 Tailwheel
60 Compressed air bottles
61 Radio compartment
62 Master compass
63 Whip antenna
64 Flap upper surface retainer
65 Flap actuating lever
66 Wing/fuselage fairing

67 Flap
68 Flap construction
69 Wing box
70 Flap
71 Balance weight
72 Trim tab
73 Wingtip
74 Navigation light
75 Pitot tube
76 Handley-Page slat
   control
77 Flap rails (two)

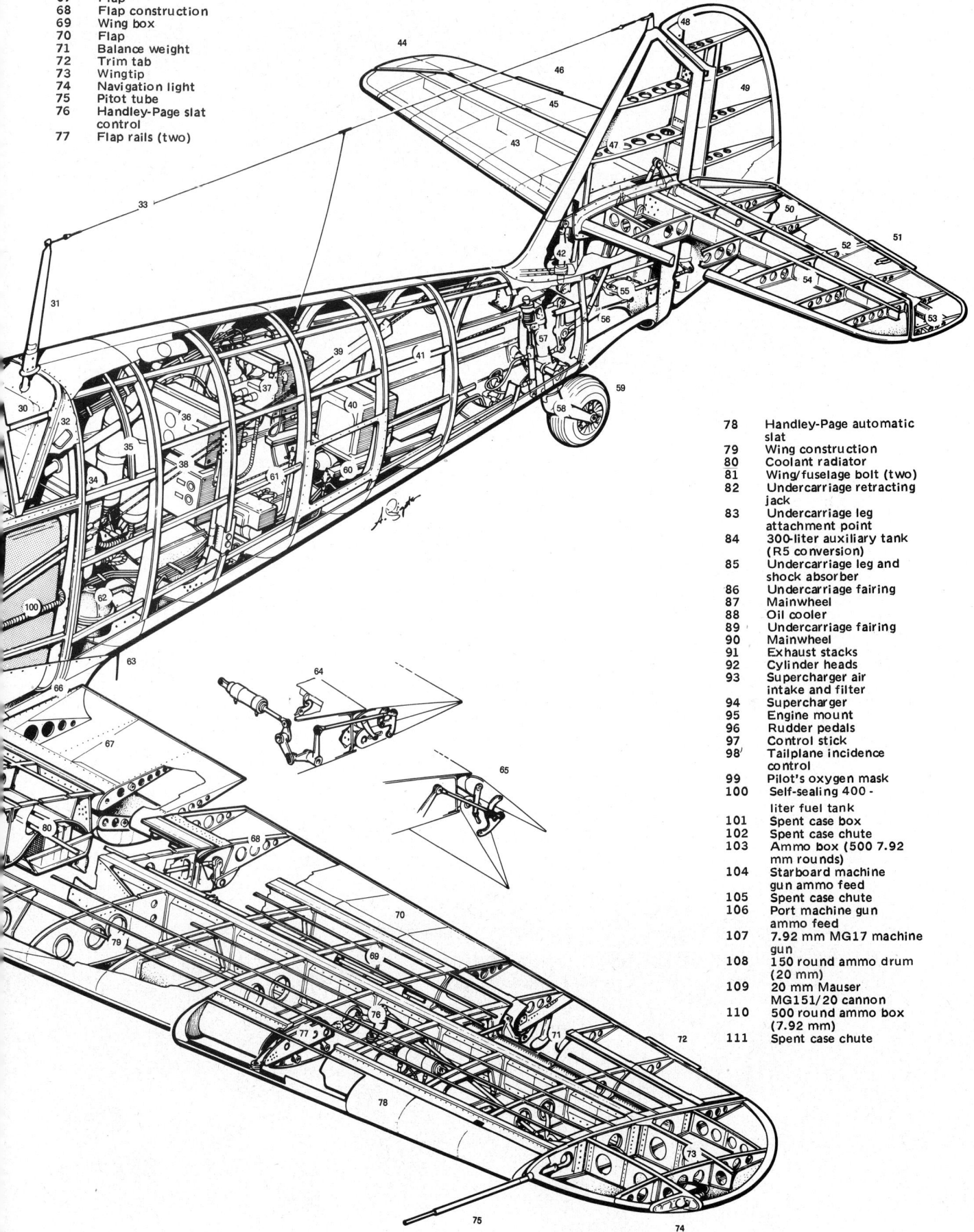

78 Handley-Page automatic
   slat
79 Wing construction
80 Coolant radiator
81 Wing/fuselage bolt (two)
82 Undercarriage retracting
   jack
83 Undercarriage leg
   attachment point
84 300-liter auxiliary tank
   (R5 conversion)
85 Undercarriage leg and
   shock absorber
86 Undercarriage fairing
87 Mainwheel
88 Oil cooler
89 Undercarriage fairing
90 Mainwheel
91 Exhaust stacks
92 Cylinder heads
93 Supercharger air
   intake and filter
94 Supercharger
95 Engine mount
96 Rudder pedals
97 Control stick
98 Tailplane incidence
   control
99 Pilot's oxygen mask
100 Self-sealing 400 -
    liter fuel tank
101 Spent case box
102 Spent case chute
103 Ammo box (500 7.92
    mm rounds)
104 Starboard machine
    gun ammo feed
105 Spent case chute
106 Port machine gun
    ammo feed
107 7.92 mm MG17 machine
    gun
108 150 round ammo drum
    (20 mm)
109 20 mm Mauser
    MG151/20 cannon
110 500 round ammo box
    (7.92 mm)
111 Spent case chute

certain number of prototypes with various weapon combinations, by 1938 the small E-O pre-production series had been finished, followed by the Bf 109E-1, complete with all its military equipment and with four guns. The two weapons in the wings could be 7.9 mm machine-guns but were more often 20 mm cannon.

Immediately, 40 were sent to Spain, but very few arrived in time to take part in the last phases of the conflict next to the surviving Bs and Cs of the

Above: a pair of Me 109F-2s belonging to 4 (F)/123. Below: a Me 109F-2/Trop of I/JG27 during testing of the landing gear operating mechanism.

Condor Legion. They all remained in Spain, and the last one survived till 1954.

The Bf 109 E-1/B was a fighter-bomber sub-type with a fuselage rack for carrying a 50 kg (110 lb) bomb.

The E series ('Emil' in German radio code, like

'Bruno' - 'Caesar' and 'Dora' for the B, C and D versions) was produced at the Regensburg works and by other sub-contracters, and by September 1939 850 had already been built.

The most common variant was the E-3, already in full production in 1939/40 at a rate of 150 planes a month. Its main feature was the improved 1175 HP DB 601Aa engine, and was intended to take the engine-mounted MG/FF cannon which, however, was hardly ever fitted because of irregular functioning. In 1940, 1868 were built, of which 340 were exported.

In the meantime, the War had spread in Europe and confirmed the expectations of the Bf 109. For the attack on Poland, the old B, C, and D were still used alongside the superior BF 109E, which maintained supremacy in the air in spite of the reactions of the Polish fighter force equipped with older PZL fighters with high wings and a fixed undercarriage.

In the Battle of France, too, the Bf 109 showed distinct superiority over the fighters of the 'Armée de l'Air' and the RAF, while in the confrontation with the RAF during the 'Battle of Britain', its superiority was counter-balanced thanks to the massive use of the Spitfire by the British forces, against which the Messerschmitt found itself fighting on equal terms and suffering, too, from a short endurance and inadequate range, a problem the British certainly did not have, as they were fighting on their own territory. Other defects of the '109': undercarriage weakness, an over-narrow wheel track, a tendency to yaw on take-off and inadequate pneumatic brakes.

The duel between the Messerschmitt 109 and the Spitfire continued the whole length of the War, in the skies and in the secret workings of the technical engineering offices in an attempt to anticipate or equal the improvements of the enemy.

For the whole of 1940, production of the Bf 109E continued, while new developments resulted in numerous variants in addition to that of the more widespread E-3, in particular the E-4 with its powerful armament of two 7.9 mm and two 20 mm guns. Other variants followed which were continually more powerful and heavier or improved in some way for other tasks, for example, photographic reconnaissance flights. Similar to the E was the T, version, a naval red version for use aboard the *Graf Zeppelin*, the only and never finished German aircraft carrier.

While these versions were being developed, a new 1350 HP DB 601E engine, with a refined cowling, a new oil cooler and a new supercharger intake was mounted on a group of 3 Emils. It also had a re-designed propeller spinner.

The struts bracing the horizontal tailplanes disappeared and the weapons were reduced to 1x20 mm and 2x7.9 mm guns. And so the new version, the Me 109F was born (F-'Friedrich'). (The Bayerische Flugzeugwerke, from July 1938, had become Messerschmitt AG., so from the version F on, we shall use the designation Me 109 instead of Bf 109). The F was considered to be the most balanced version, the optimum obtainable from the original airframe. Among other modifications, the fuselage was lengthened, the wing-tips were rounded and the main landing-gear legs were broadened out to obtain a wider track.

The Me 109F enjoyed a brief period of superiority over the Spitfire, forcing the British to hurriedly 'runfor cover'. There was a huge number of variants of the 'Friedrich', but almost standard features were the 3 fuselage guns, with only the Me 109F-4/R1 carrying 2 wing cannon of 20 mm. From the Me 109 F-2 on, the Oerlikon MG/FFs were replaced by the more efficient and modern Mauser MG151s. There were many operational models of the '109' but the 'F' was probably the one with the most experimental changes, often with interesting results like the Me 109Z (Zwilling), a 'Siamese twinning' of two Me 109F-1s. One Me 109F-4 was given a 'butterfly' tail assembly; others tried out tricycle landing gear of different configurations, accumulating useful experience in preparation for the future Me 309. Antislip ailerons, high-aspect ratio wings for high altitudes, BMW 801 radial engines (for comparison with the

---

Messerschmitt Me 109F-4/B Trop. (Friedrich Jabo)
Day fighter-bomber, single seat

Power plant: one Daimler-Benz DB 601E-1, 12-cylinder, inverted Vee, liquid-cooled engine, rated at 1350 HP for take-off (2700 rpm) and 1300 HP at 5,500 m. Prop: three-blade variable pitch, constant speed VDM of 2.96 m diameter. Fuel capacity: 400 liters.

Dimensions: wing span: 9.92 m; length: 8.95 m; height: 2.60 m (top of the radio mast); wing area: 16.20 sq. m; undercarriage track: 2.20 m.

Weights: empty (F-4): 2,386 kg; empty, equipped (F-4B): 2,785 kg, (F-4B/Trop.): 2,795 kg; loaded (F-4B): 3,005 kg, (F-4B/Trop.): 3,015 kg; max. overload (F-4/R1): 3,117 kg.

Performance: max. speed: 545 km/h at 6,500 m; initial climb rate: 22.10 m/sec; climb to 1,000 m (standard F-4): 54 sec, to 3,000 m: 2 min 36 sec, to 5,000 m: 5 min 12 sec; service ceiling: 11,000 m; combat radius: 225 km; range: 500 km, with 300-liter external tank: 750 km.

Armament: two Rheinmetall-Borsig MG17 7.92 mm machine guns firing thru the propeller disc, with 500 r.p.g. and one Mauser MG151/20 20-mm cannon (engine mounted) with 150 rounds, plus (F-4B) one 250 kg SC250 or one 500 kg SC500/SD500 bomb.

Fw 190), DB 603 engines with an annular radiator and the pressure system of the Me 309, were examples of the equipment tests using the Me 109F as a flying test-bed. In the meantime, German military involvement had spread to the Balkans, the Mediterranean and Russia, all sectors in which the Messerschmitt once again confirmed its high quality (far superior to that of its enemies) and exciting the envy of the Italian pilots, fighting alongside the Germans in a heterogeneous mass of obsolete fighters. In the campaign against Yugoslavia, there were even some encounters between Messerschmitts of opposite factions, the Yugoslav air force having 73 Bf 109E-3s in service.

Certain aces became famous with their 'Friedrich', including (in the Mediterranean), Walter Oeseau and Hans Joaquim Marseille; the latter achieved the highest number of his 158 victories in his personal F-4 'Gelb (yellow) 14'. Strong ties often grew between the Me 109 and its pilots, which were maintained even after better planes had been brought into service. It was not unusual to find 'aces' remaining faithful to their 109s even when the rest of the squadron under their command was using the Fw 190. These ties were found among the last versions of the Me 109 which, although more powerful, heavier and faster, did not have the overall ease of handling of the 'Friedrich'.

In the Summer of 1942, the Me 109F was replaced on the production lines by the Me 109G (Gustav). It was by far the most important version, comprising almost 70% of the total production of 33-35,000 Me-109 aircraft.

The 'raison d'être' of the new model was the adoption of the DB 605A engine of 1475 HP; otherwise there was little difference between the 'Friedrich' and the early 'Gustav', and the transfer from one to the other was entirely 'painless'. The operational ceiling of the DB 605 was much higher than that of the previous models and in the G-O pre-production series and the G-1, the cockpit was pressurized. From the Me 109G-5 variant on, the 7.9 mm synchronized MG17s were replaced by 13 mm MG 131s.

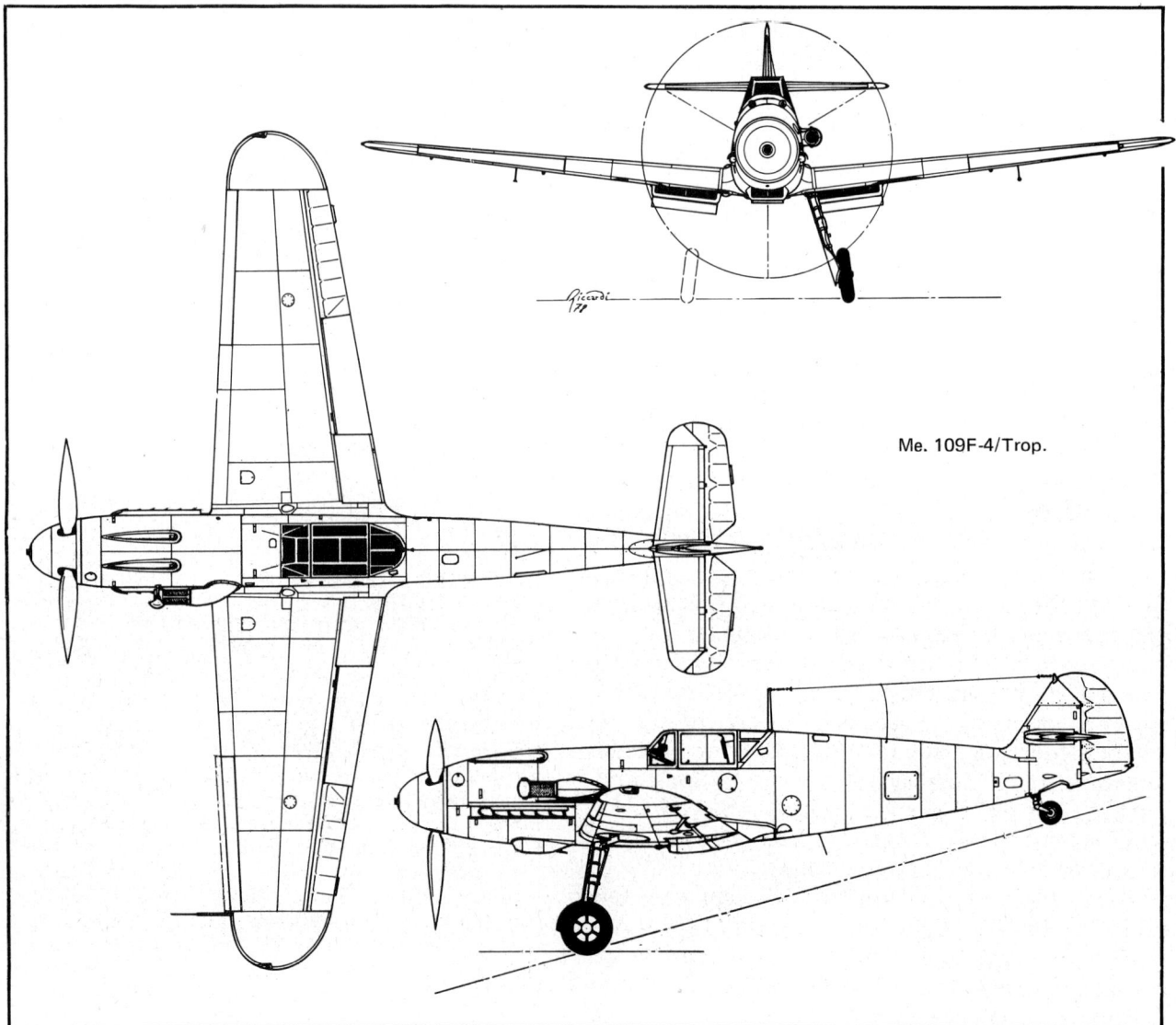

Me. 109F-4/Trop.

An unusual view of a Me 109F-2/Trop photographed during a low level pass.

The subsequent variations, especially the G-6 with numerous modifications, adopted progressively heavier armament, building up an increasing spiral which involved the power of the engine, structural strengthening and weight. In particular, the engine-mounted MG151 was replaced by the 30 mm MK108 on an increasing number of models. On different sub-variants, 2x20 mm or 30 mm cannon were mounted in underwing gondolas, and on the G-6/R2, 2x2 cm mortars were mounted. Production of the Me 109G continued virtually till the end of the War, the last series differing from the first in various aesthetic points: above all, the form of the vertical tail-plane had been changed and a canopy giving improved visibility had been adopted, known as the 'Erla Hood' or (more incorrectly) as the 'Galland Hood'.

After the G-1/Trop., only the models with uneven numbers (G-3 and G-5) were pressurized. None of the subsequent variants of the 'Gustav' were pressurized, and thus were given even numbers: G-6, G-8, G-10, G-12, and G-14. The design of the fuselage was modified with the last of the DB 605 engines; the new cowling appeared for the first time on the Me 109G-6/AS with the DB605AS engine using the DB603 supercharger, and the design also enclosed the huge bulk of the MG131 cradles. Along with the new streamlined cowling, the new vertical tailplane section become standard too. With these new modifications to the G-6, G-10 and G-14, the look of the Me 109 became definitely more modern and aggressive, but this was not enough to stop the incessant air offensive which was devastating Reich territory: the days when the Luftwaffe darkened the skies were certainly over, and even its claims to high quality became questionable when compared to the Allies later production aircraft.

After 1942 began a gradual withdrawal of the German armed forces into the heart of the Reich in an increasingly defensive attitude. Africa had been lost, Italy was partly in the hands of the Americans and British, the industrial potential of the immense Soviet Union, (helped by continuous Allied supplies) was now capable of counter-attacking the German advances and a new

front was opening in the West. German losses had become serious: the best pilots, those who after Spain had continued the fight in the skies of

Above: the production line for the Me 109G. Below: a Me 109G-6/R3 with a ventral 300 liter tank. Both on the planes under construction and on those seen taxying in the photo can be seen the D/F loop antenna. This was assembled in the factory on the last subversions of the Me 109G.

Europe, were, for the most part, dead, and training of new pilots was insufficient, so much so that they were often shot down in their first combat, because, too, of inferiority in numbers. The Luftwaffe, in its last days, had in fact the unusual feature of having side by side a high percentage of pilots destined not to survive their first battle and a strong nucleus of super-aces with well over .100 victories (352 in the case of Erich Hartmann).

This was perhaps due also to the fact that the Me 109G was the first 'less simple' version of the Me 109, and this led to the development of a two-seater trainer version called G-12.

The last versions of the Me 109G were subjected to such an extraordinary number of modifications that it was difficult to keep up with them, and after the appearance of the G-10 and G-14, there was still one last version. In fact, the G-10 was an interim model with various modifications which were standardised on the Me 109K ('Kurfurst'), the last model before the end.

The particular moment in history, the lack of precise data, and the different interpretions of aeronautical historians, have often made it impossible to distinguish between the 'Kurfurst' and the

Me 109Ga-5 (former Trop). Magyar Kiralyi Legiero.

1 Vadasz Osztaly 'Puma'. Hungary 1944.

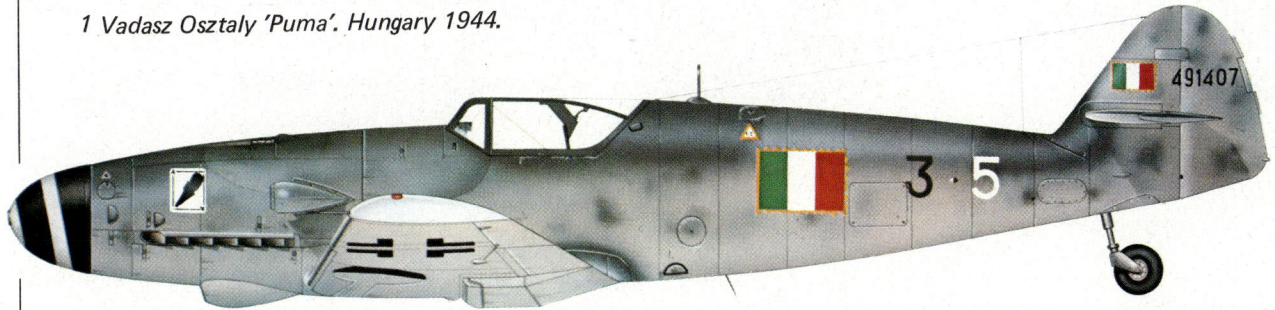

Me 109G-6AS with a DB-605ASC engine. Aeronautica Nazionale Repubblicana. 1st Gr. CT 'Asso di Bastoni', 3.a Sq. Italy late January 1945.

Me 109K-4. Luftwaffe, I/JG77 'Herzas'. Western front. Winter 1944-45.

Me 109G-6/Trop. Luftwaffe, 7/JG 27.

Me 109G-6. Luftwaffe. Eastern front.

drawings by Marco Gueli

From top to bottom: a Me 109G (perhaps a G-5 or -6) abandoned as unserviceable. Another Me 109G (in the series included between the G-4 and the G-8) with R7 aerial during testing. One of the very first Me 109G-6/Trop of II/JG51 shot down in Sicily in the Summer of '43.

series in production at the same time. The Me 109K has always been a rather mysterious 'object'.

After the K-0 pre-series, a result of modifications to some G-5s, production was fixed on the K-4 version which first appeared in October 1944. The Me 109K clearly revealed the effort to adapt the design to the needs; in fact, range was limited, long flights for reaching the objective no longer being necessary. Speed hovered near 730 km/h (450 mph), fire-power was enormous, reaching 3x30 mm cannon in some cases. They were real 'monsters', following a pattern which for some years had become what could be described as regressive.

Together with other fighters of the Luftwaffe, the 'Gustav' and the 'Kurfurst' took part in the last important offensive operation: operation 'Bodenplatte' put into action in the early hours of New Year's Day 1945. 1000 German fighters attacked the Allied positions, achieving impressive

'Gustav', and it is probable that in the last period they built real 'harlequins' of the air, the result of putting together components of the various

results, but losing 200 planes and, more important, many precious pilots. Then came the end for the Luftwaffe and the Me 109, which had been its companion for 7 years of bloody battle. Like all successful planes, the Me 109 was widely exported, to practically all the air forces allied with the Luftwaffe. But not only to these: substantial supplies were sent to neutral countries such as Switzerland and Spain, primarily for political reasons, but also because of the need for hard currency.

The Bf 109E was delivered to, besides Spain (which had kept models received just before the end of the Civil War): Switzerland, Yugoslavia, Bulgaria, Croatia, Czechoslovakia and Rumania, plus three evaluation models to Japan. Even the Regia Aeronautica came into possession of an 'Emil', probably an ex-Yugoslav model. The Me 109F was less widely exported, only Hungary receiving a substantial quantity. A few others went to Spain and a very few (perhaps only two) to Italy.

The 'Gustav', on the other hand, was used in great numbers under other names: in Finland in particular, then Hungary, Bulgaria, Spain, Croatia, Switzerland, Czechoslovakia and Italy, where it was in service both with the Regia Aeronautica, before the Italian Armistice, and the Aeronautica Nazionale Repubblicana, which continued to fight beside the German till the end of the War. A few Me 109Ks, in the K-4 version, were delivered, according to documents of the time, to the Italian

The cockpit and canopy of a Me 109G-10. The red line painted on the side front window (seen in the photo across the pilot's face) is almost invisible. These were used for reference during dive-bomb attacks.

Above: a Me 109G-10/R3 with a DB-605D engine and Fo 987 oil cooler. Under the left wing is the aerial for FuG16ZY. Below: a Me 109G-10 of the Croatian Airforce.

A Hauptmann prepares to climb into his Me 109G-3, which carries the insignia of the Gruppe Kommandeur.

ANR, but there are good reasons to believe that they were hybrid versions of the G-10/K-4.

Me 109s were produced under licence in Rumania, Hungary, Switzerland and Spain. The history of the Me 109 is far from ended. Production continued uninterrupted for a further 15 years after the end of the War! In Czechoslovakia the Avia firm of Praga-Cakovice had been involved in the production program of

the German fighter during the War, and in the Spring of 1945, was beginning to make the first deliveries of a batch of Me 109 G-14s. Production started up again after the War with the aim of keeping the factories busy, of supplying the new Czech air force with fighter planes and above all for export. A bomb attack, however, destroyed

Above: a Me 109G-6/Trop, Werk Nr16416 'Weiss 9', captured and taken to America. Below: a line of Me 109G-1s ready for delivery.

the stock of DB 605 engines and only 20 C10 single seaters (Me 109 G-14) were completed, together with two models of a twin-seat version

designated C110. The designations were then changed to S-99 and CS-99 respectively. Production continued with the only engine at their disposal at the time, the Jumo 211F of 1350 HP, but with an inadequate performance for fighter planes; the result was a hybrid which was not greatly appreciated by its pilots, who nicknamed it 'Mezec' (Mule). It was particularly dangerous on take-off and landing. Conceived as the S-199 (CS-199 was the twin-seat version) it was offered for export as the C210 and became the first fighter plane in the Israeli Chel Ha'avir. It was thus that in the Sinai, the duel between the Me 109 and the Spitfire was repeated, this time between Israelis and Egyptians; the latter also had in service a batch of Italian Macchi C.205 fighters, also of 1940-45 vintage. The last Czech S-199s were withdrawn from service at the end of the 50s. At that moment, however, in another part of Europe, the Me 109 was still in production! In fact, Spain had continued uninterrupted production of the German fighter, in their case too having recourse to other engines than the DB 605, no longer available. Hispano Aviacion began in 1945 by equipping 25 German produced Me 109G-2 airframes, with the Hispano Suiza 12Z-89 1300 HP engine, and the resulting plane was named the HA-1109-JIL; then, on entirely Spanish made airframes, they mounted the superior HS 12Z-17, also of 1300 HP. This new fighter was called HA-1109-KIL, available from 1951. Other versions with the same engine followed, until they decided on the final model, using Rolls Royce Merlin 500-45s of 1400 HP. The plane came into existence as the HA-1109-MIL but the final version was called HA-1112-MIL and was widely used by the Ejercito del Aire as a close-support fighter, carrying 2 Hispano HS 404 or HS 804 cannon and 8x80 mm Oerlikon air-to-surface rockets.

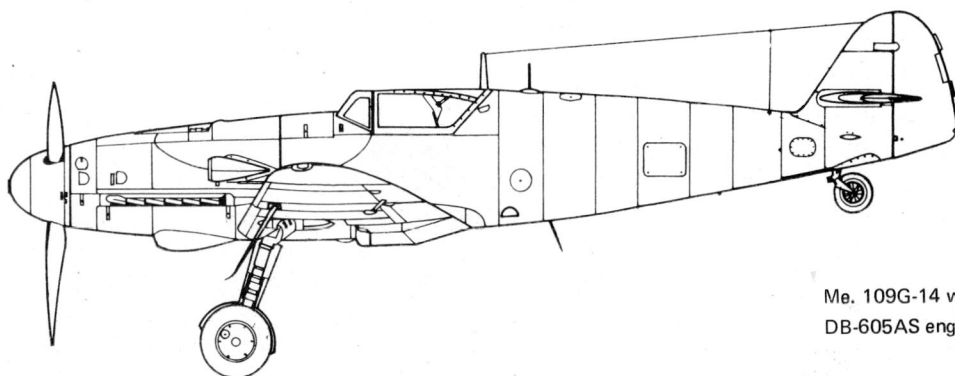

Me. 109G-14 with DB-605AS engine

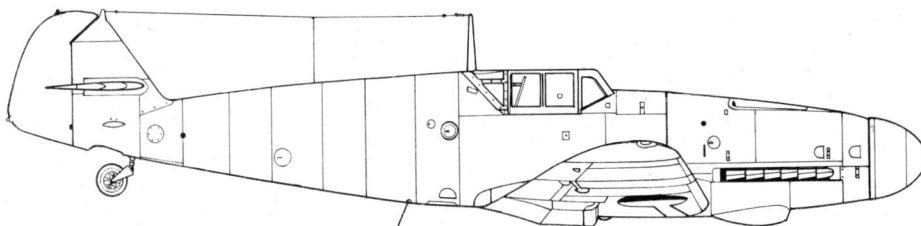

Me. 109G-2

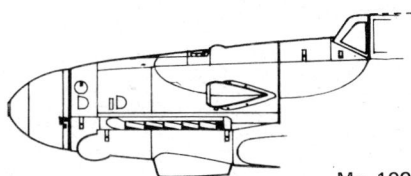

Me. 109G-10 with DB-605D engine

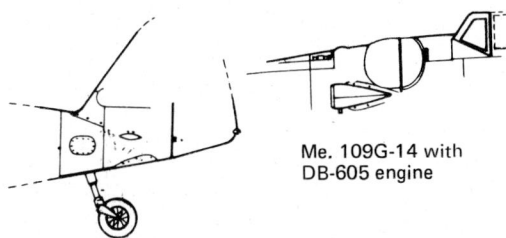

Me. 109G-14 with DB-605 engine

drawings by G. Valentini

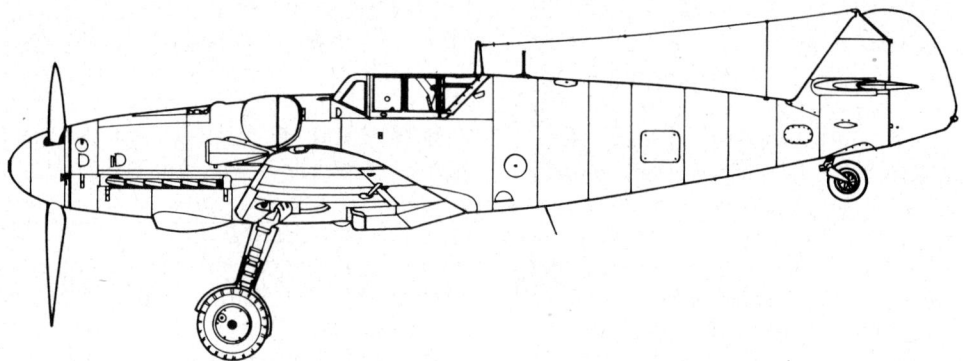

Me. 109G-6

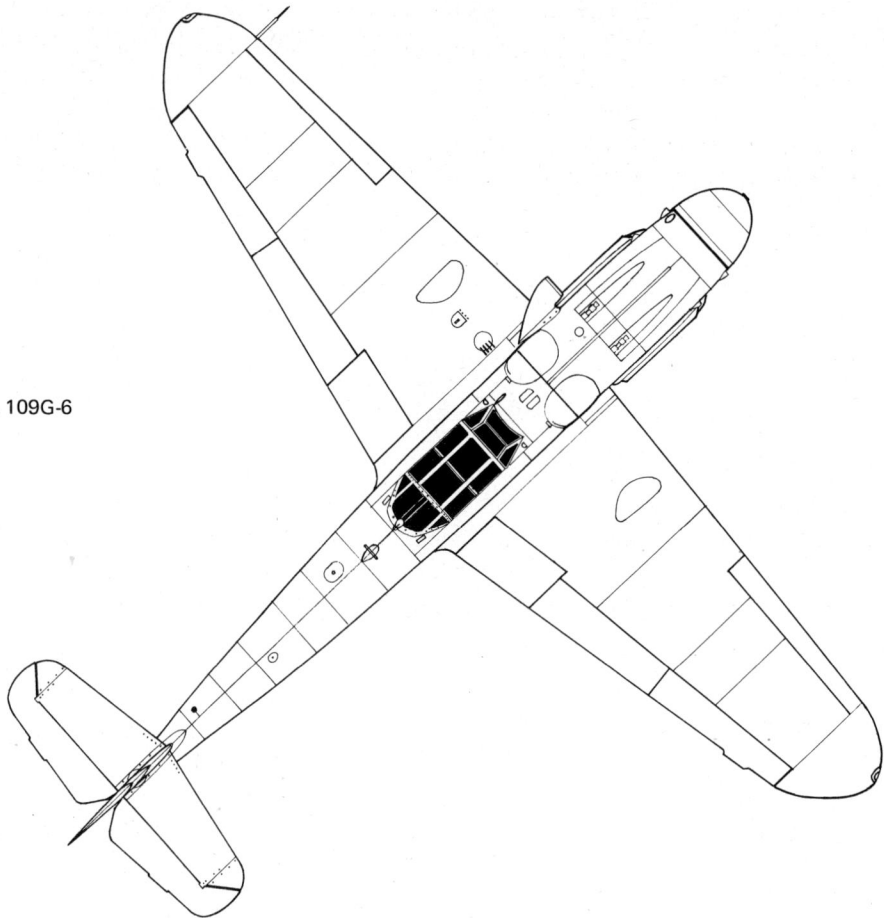

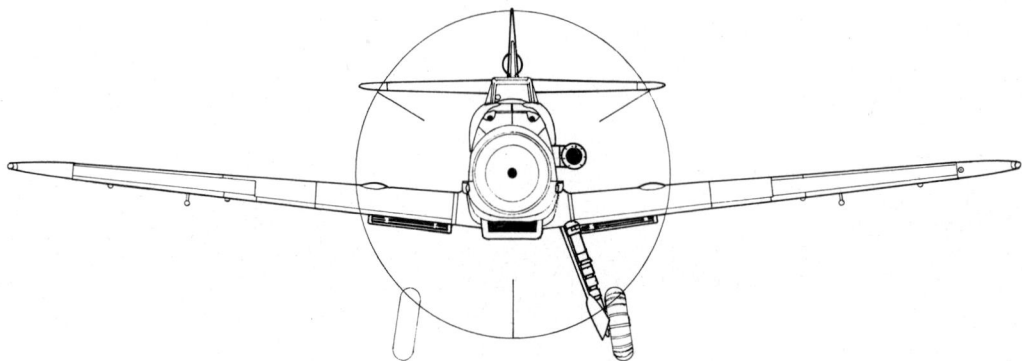

drawings by G. Valentini

MESSERSCHMITT Bf 110 'Zerstörer'

A classic view of a pre-production Bf 110C-0.

This first, and most typical, representative of the multi-purpose twin-engine class of fighter, playing an important role in the airborne operations of World War II, originated in a specification issued by the RLM in 1934 for a strategic twin-engined fighter, whose service classification, that of 'Zerstörer' (destroyer), was taken from naval terminology.

The design of the Bf 110 presented many structural similarities to the slightly earlier Bf 109 single-engined plane, and an exceptional performance was expected of it thanks to the new Daimler-Benz engines. The first prototype flew on 12 May 1936 with two DB 600A engines. By the end of the year two other prototypes had joined it, and tests revealed that the plane was up to expectations. However, the DB 600 engines were not fully developed, and the promising DB 601 engines were not ready in the time estimated. In spite of this, the service testing was made with the less powerful Junkers Jumo 210 of 680 HP, which went on to equip the four A-0 pre-production models, very similar in appearance to the prototypes. These were followed by 10 Bf 110B-0s with the more powerful Jumo 210G 700 HP engines, and about 35 B-1, B-2, B-3 mixed; all differed from the 'A' series by their cleaner fuselage nose, which remained a feature of the subsequent versions.

Armament varied from a minimum of 4x7.9 mm to a formidable maximum of 4x7.9 mm + 2x20 mm guns; in addition, on all models there

Close-up of a pair of Bf 110D-2s of III/ZG 26 'Horst Wessel' while flying over a flottilla of fishing boats in the Aegean Sea.

A Bf 110C-7 of the Schlachtgruppe I during the 'Redbeard Operation' on the Eastern front.

Messerschmitt Bf 110C-3 Zerstoerer cutaway drawing key

1. Nose fairing
2. Upper battery: four 7.92 mm MG17 machine guns
3. Machine-gun nozzles
4. Ammo feed chutes
5. Ammo boxes (1,000 rpg)
6. Lower battery: cannon blast tubes
7. Mainwheel
8. Exhaust stacks
9. Oil cooler air intake
10. Engine instrument gauges

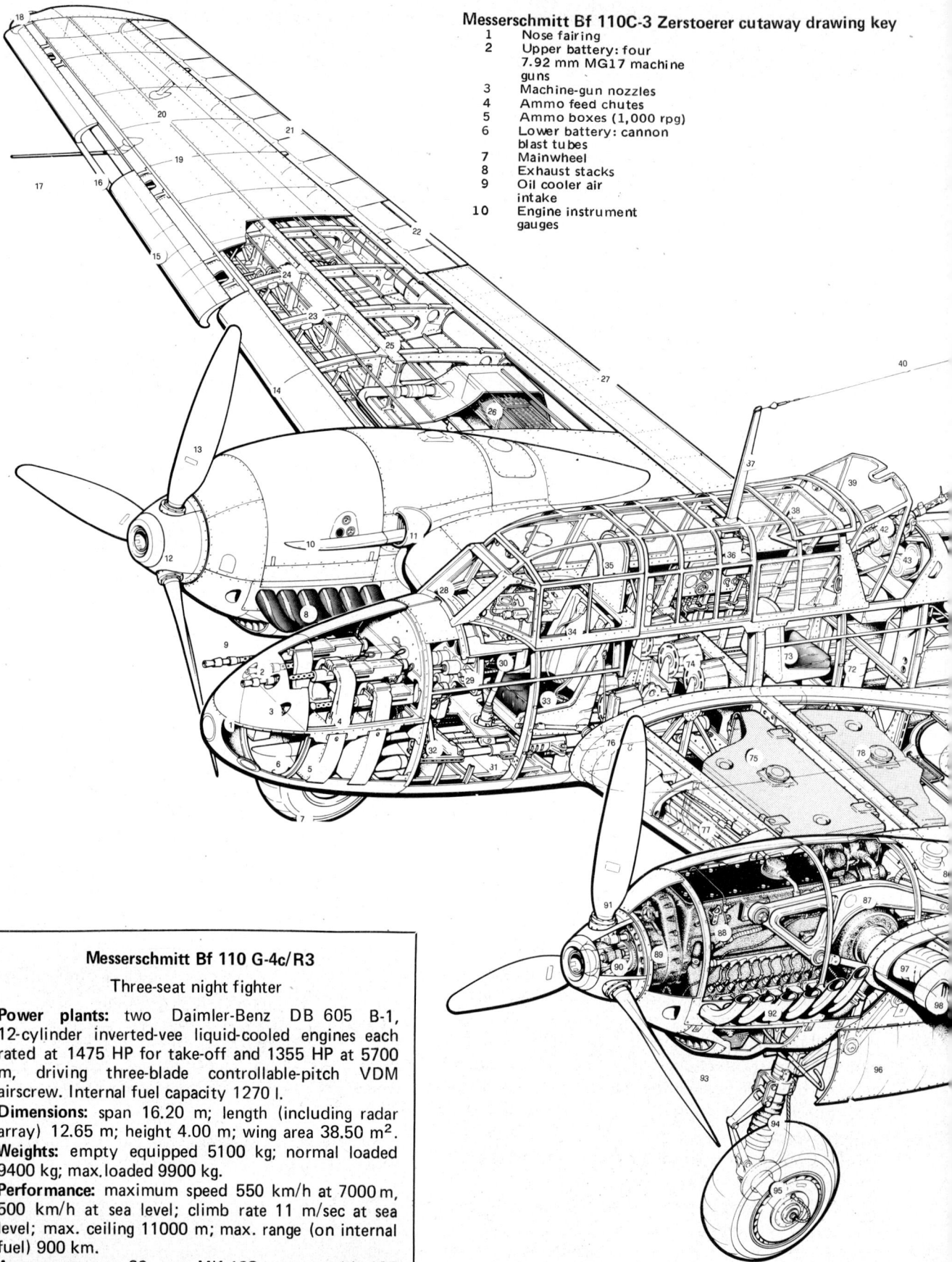

Messerschmitt Bf 110 G-4c/R3

Three-seat night fighter

Power plants: two Daimler-Benz DB 605 B-1, 12-cylinder inverted-vee liquid-cooled engines each rated at 1475 HP for take-off and 1355 HP at 5700 m, driving three-blade controllable-pitch VDM airscrew. Internal fuel capacity 1270 l.

Dimensions: span 16.20 m; length (including radar array) 12.65 m; height 4.00 m; wing area 38.50 m².

Weights: empty equipped 5100 kg; normal loaded 9400 kg; max. loaded 9900 kg.

Performance: maximum speed 550 km/h at 7000 m, 500 km/h at sea level; climb rate 11 m/sec at sea level; max. ceiling 11000 m; max. range (on internal fuel) 900 km.

Armament: two 30 mm MK 108 cannon with 135 rpg; two 20 mm MG 151 cannon, in the ventral tray, with 300 rounds (port) and 350 rounds (starboard), and two similar weapons (or MG/FF 20 mm cannon) in a 'schräge musik' installation.

11	Supercharger air intake
12	Spinner
13	VDM metal three-blade propeller
14	Wing leading edge
15	Handley-Page automatic slat
16	Wing leading edge construction
17	Pitot tube
18	Navigation light
19	Wing skinning
20	Aileron balance weight
21	Aileron
22	Trim tab
23	Wing ribs
24	Control runs
25	Wing spar
26	Coolent radiator
27	Flap
28	Armored windshield (57 mm)
29	Instrument panel
30	Control stick
31	Gun barrels
32	Rudder pedals
33	Pilot seat
34	Safety harnesses
35	Armored headrest
36	Aerial tuner
37	Aerial mast
38	Top glazing (open)
39	Rear glazing (open)
40	Dipole aerial
41	Dipole retainer
42	7.92 mm MG15 machine gun
43	7.92 mm ammo (750 rounds)
44	Boarding handle
45	FuG10 transceiver (from Bf110C-2 onward)
46	Oxygen bottles
47	Fuselage skinning
49	Fuselage construction
49	Compass
50	Control runs
51	Tailwheel frame
52	Tailwheel leg
53	Tailwheel
54	Tailplane
55	Fin
56	Balance horn
57	Rudder
58	Trim tab
59	Elevator
60	Trim tab
61	Elevator construction
62	Trim tab

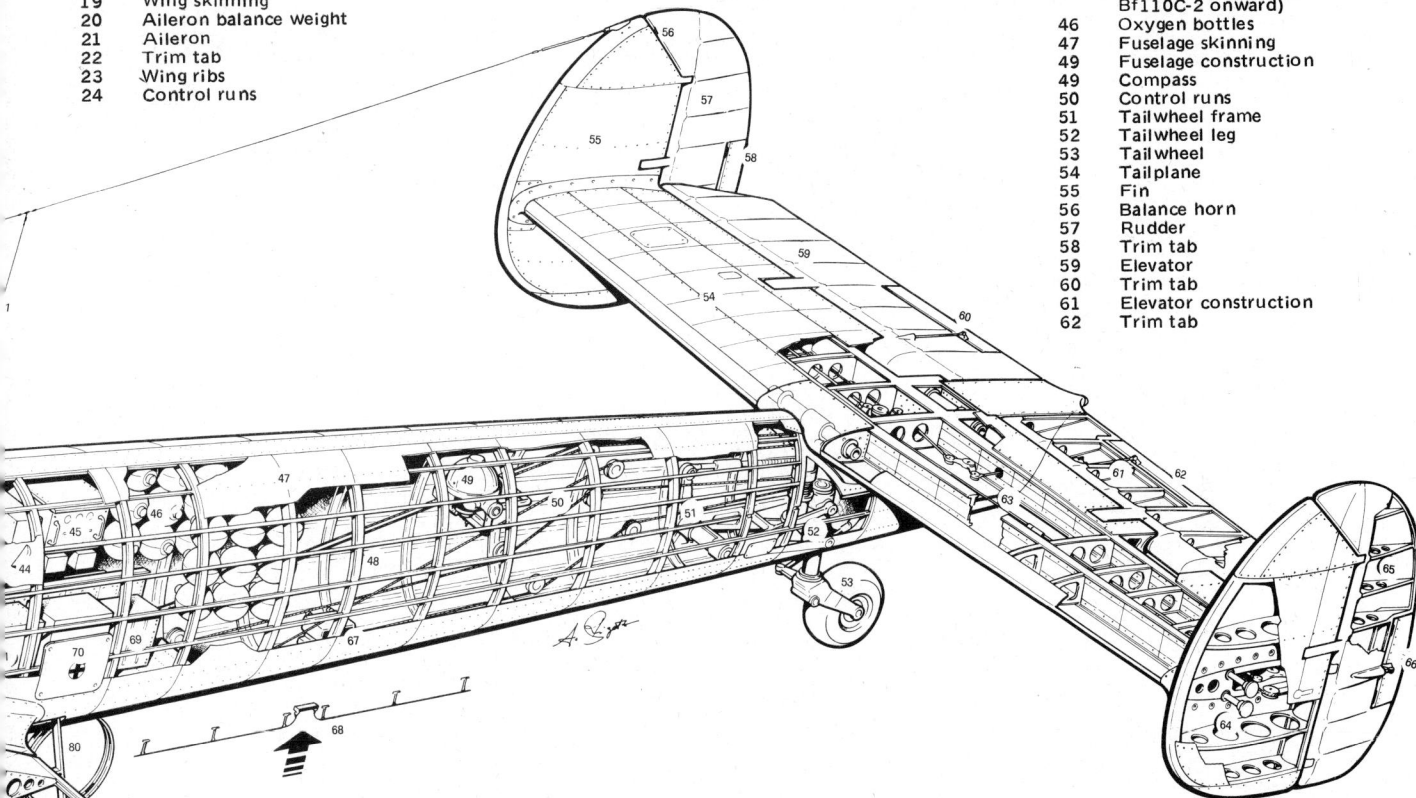

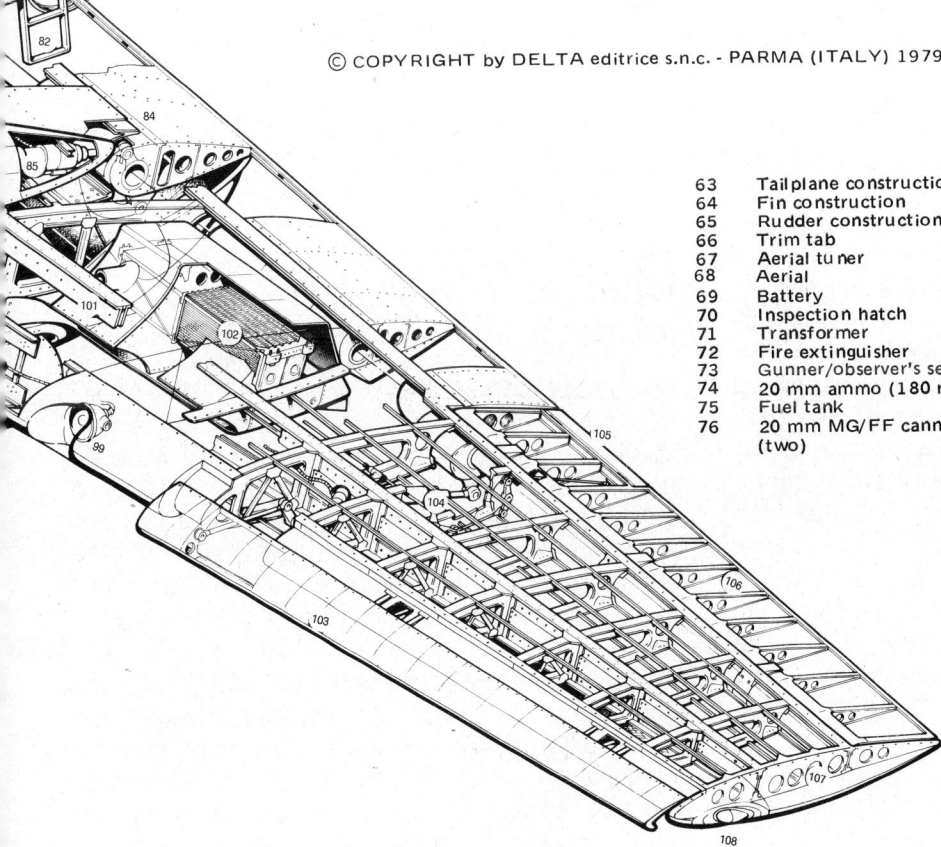

63	Tailplane construction
64	Fin construction
65	Rudder construction
66	Trim tab
67	Aerial tuner
68	Aerial
69	Battery
70	Inspection hatch
71	Transformer
72	Fire extinguisher
73	Gunner/observer's seat
74	20 mm ammo (180 rounds)
75	Fuel tank
76	20 mm MG/FF cannons (two)
77	Wing leading edge construction
78	Fuel tank
79	Flap rib
80	D/F loop
81	Wing/fuselage fillet
82	Retractable ladder
83	Tube spar
84	Flap
85	Flap control
86	Oil tank
87	Engine mount
88	Daimler-Benz DB 601A-1 engine
89	Reduction gear
90	Propeller pitch gear
91	VDM three-blade metal propeller
92	Exhaust stacks
93	Oil cooler air intake
94	Shock absorber
95	Mainwheel
96	Landing gear door
97	Supercharger air duct
98	Supercharger air intake.
99	Landing light
100	Wing front spar
101	Wing center spar
102	Coolant radiator
103	Handley-Page automatic slot
104	Aileron control rod
105	Trim tab
106	Aileron construction
107	Wingtip
108	Navigation light

A pair of Bf 110D-3s of ZG 26 with 900 liter reserve tanks.

Another Bf 110D-3, probably belonging to III/ZG 76, in action in the Mediterranean in 1941.

was a 7.9 mm gun on a flexible mount for rear defense.

This fine Messerschmitt twin-engine plane reached full development in the subsequent version, the Bf 110C, thanks to the adoption of the excellent DB 601A engine and a few small modifications, fruits of the experience already gathered from the previous models. After the usual 10 pre-production C-0 models, manufacture and delivery of the production models, C-1, C-2, and so on, were rapidly begun, characterised by subsequent improvements to equipment and weapons. At the beginning of the War, a few hundred models of the Bf 110 were already in the Luftwaffe's inventory and production was being expanded.

Poland, Norway and France were the first testing grounds for the powerful 'destroyer', and seemed to confirm the optimistic expectations of its supporters. The situation changed radically with the Battle of Britain when the presence of an opposition with an aggressive, well-trained fighter force equipped with excellent Spitfires and Hurricanes placed the Bf 110 in a critical position, especially as it had been designed to compete directly with traditional fighters. The losses inflicted on the 'Zerstörergruppen' (to which were added the latest versions of the C and D models) were heavy, and this forced the Germans to reconsider the effectiveness of the twin-engine airplane as a strategic fighter.

In connection with this, it is interesting to note that another specialty aircraft, the 'Sturzkampfflugzeug' (the 'Stuka'- dive bomber) also found itself in a crisis during the Battle of Britain, in spite of recent victories. This, with the 'Zerstörer' was to have constituted the triumphant 'pair' of the Luftwaffe, or so its chiefs (had hoped).

From this moment on, the Bf 110 and its successors were used with excellent results in more suitable roles such as ground attack, high-speed reconnaissance, night flying and long range day flying against unescorted bombers, thus exploiting its excellent characteristics of speed, firepower and endurance.

The Bf 110D was a long range version, found necessary for the operation against Norway. In the Spring of 1940 the D-0s were delivered, recognizable by their enormous ventral fuel tanks holding 1050 liters (264 Imp. gals.) and theoretically jettisonable in flight. This installation proved to the extremely dangerous and after producing a few of the D-1 series thus equipped, the Germans passed on as quickly as possible to the D-1/R2 version with two supplementary tanks attached to the wings, each containing 900 liters (198 Imp. gals.). Development of the D versions continued with modifications for night flights, convoy escort and other uses.

The availability of more powerful engines led to a multiplication of new versions of the Bf 110, particularly the E version with 1200 HP DB 601N engines, and later the F with 1300 HP DB 601F engines. Previously the DB 601N had been mounted on fighter-bomber sub-variants of the C version.

At the same time, German involvement was extending like an 'oil slick' over the Balkans, the Mediterranean, in Africa, briefly in Iraq, and, in the Summer of 1941, into the USSR.

However, the myth of the Blitzkrieg was faded for good and victory would be sought with greater and greater difficulty, using increasingly complex, sophisticated and even bizarre arms and equipment. An airplane like the Bf 110 was an ideal test and operational platform for reserve tanks of every kind, for rocket projectiles, large-caliber cannon, oblique upward firing guns ('Schräge Musik'), infra-red sensors for interception of targets at night, radar, and a whole variety of offensive loads. The result was an incredible number of variants on the original model.

Production, after reaching a maximum in '40 and '41, began to drop and was halted altogether eventually, making way for the new, more modern Me 210, on which the RLM had now placed its greatest hopes. In fact, the inadequacy of the Bf 110's performance was becoming more and

A Bf 110 in North Africa.

more obvious and the new, more complex tasks demanded of it brought out the limitations of the original design. Unfortunately, the Me 210 proved to be so full of defects that it was taken out of production even before first deliveries to the operational squadrons of the Luftwaffe.

Hurriedly, the Bf 110 was resurrected and put back on the production line to fill the 'void' left by its rather undistinguished successor. New production started up again in 1942 and comprised both the last series of the F version and the new G version, with its more powerful 1475 HP DB 605 engines. With the added power, structural reinforcements had to be made to enable the old airframe to stand up to the new demands on it. Rear defense armament was increased to 2x7.9 mm guns, while from the G-2 variant on, the two 20 mm MG/FF cannon were replaced by the better MG151 of the same caliber.

With the Bf 110G, too, there was an enormous

The Bf 110Cs were used as night fighters in the Regia Aeronautica Italiana, too. In the photo, a plane of the 235.a Sq.

proliferation of variants, as with its predecessors, but this model found a more successful rôle as a night fighter, thanks to the new airborne interception radar perfected by Telefunken in the early days of the RAF air offensive against the Reich. The FuG 212 'Lichtenstein' C-1 radar system (a mass produced version of the earlier FuG 202 'Lichtenstein' BC) had entered into service in the Summer of 1942, and equipped large numbers of Bf 110G-4, a 3-seat night fighter, given the unavailability of more suitable planes (like the specially designed Ju 88) considered ideal for the rôle. The Bf 110, however, proved to be a more than acceptable compromise, and remained for a long time the back-bone of the German night fighter force.

In the second half of 1943, the German night fighter force adopted the FuG 220 'Lichtenstein' SN2, operating on a wavelength of 490 megacycles as a result of British bomber countermeasures against the FuG 212 operating on 90 megacycles, the same frequency as the

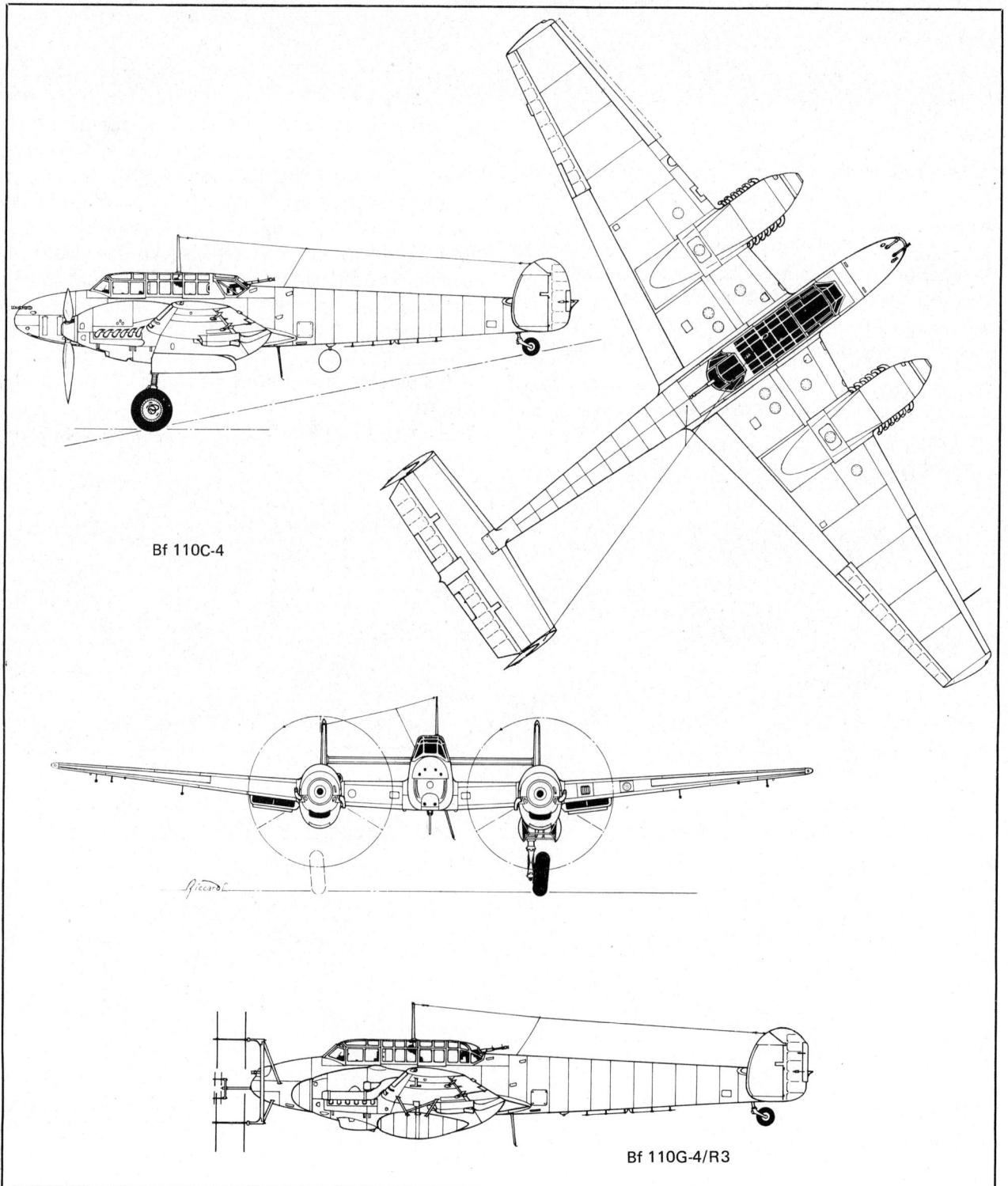

Bf 110C-4

Bf 110G-4/R3

'Wurzburg' early warning system which operated from the ground.

Gradually, production of the Bf 110 was run down, until in 1945 it was halted altogether. More suitable night fighters had entered service and were now available in sufficient quantities. A final version of the Bf 110 had been put on the assembly line next to the Bf 110G some time previously; it was known as the Bf 110H and had DB 605E engines, some structural reinforcement and a retractable tail-wheel.

Towards the end of the War, with the gradual withdrawal of the Bf 110 as a night fighter (though some Nachtjäger Geschwader still had them in service at the end of the War), the old twin-engine was given new but equally important if 'costly' tasks, such as ground attacks or heavy day-time attacks: a return to its origins, but in much more dramatic circumstances and without 'dominion of the air' any more.

More than 6000 examples of the Messerschmitt Bf 110 were built between 1936 and 1945. This airplane did not come up to early expectations and did not prove to be the invulnerable plane Goering had been counting on, but its presence in the German War inventory turned out to be a happy circumstance for the Luftwaffe and in its adaptability justified the pride and confidence which German aeronautical engineering had known in the mid-thirties.

The Bf 110 was certainly by far the best of the twin-engined combat aircraft of its class which appeared in the years immediately preceding the War. There were also the Potez 63, the Fokker G-1, the PZL P.38 'Wilk' and the Breda 88. The credit for its enormous success must go, however, to the availability of powerful modern engines like those from Daimler-Benz. Exports of the Bf 110 were insignificant and mostly concerned allies of the Luftwaffe: a few night fighters went to Rumania, not more than 6 to Hungary, and only 3 to Italy. Of the three planes used by the Regia Aeronautica, two were Bf 110Cs, to which was added a Bf 110 G-4 fitted with FuG 212. This was destroyed in an air accident and the other two, after the Italian Armistice, were transferred to a night fighter squadron of the RSI (Repubblica Sociale Italiana) together with two other Bf 110Gs given by the Luftwaffe and taken back after the alliance broke up.

MESSERSCHMITT Me 210/410 'Hornisse'

The first prototypes of the Bf 110 were still at the experimental trials stage when the RLM ordered from Messerschmitt a successor to the promising first 'Zerstörer' on which so many hopes for the revived Luftwaffe were placed. The new plane was to be more brilliant and more versatile than its predecessor, having to fulfil equally well the rôles of heavy fighter, dive bomber, attack and reconnaissance plane. The extraordinary innovation of the design of this new plane was the adoption of defensive weapons in remotely controlled electrically operated barbettes on the sides of the fuselage.

The offensive weapons were housed in an unusual way in a formed metal bomb bay beneath the cockpit, giving the nose rather a hypertrophic appearance in comparison with the comparatively slender tail section. Such was the confidence of the RLM in the quality of the Me 210 that they ordered 1000 models before the first prototype had even flown. The Me 210V1 with two 1050 HP DB601As flew for the first time on 2 September 1939, immediately revealing extremely serious defects in stability, as well as several other aerodynamic weaknesses. The first attempt at finding a solution was the replacement of the twin fins and rudders with a single tail assembly, but the improvement was marginal and the situation remained, incredibly, the same for more than two years, in spite of the continuous trials which the prototypes had to undergo.

Apart from the first three experimental prototypes another 15 planes from the A-0 series were given the name 'Versuchs' (experimental) and were used for testing operational equipment intended for the production models which the RLM kept pressing for. On the other hand, the number of defects discovered was so high that to eliminate them would have prevented the company from fulfilling production promises within the time agreed upon. Delivery of the first Me 210A-0s (94 in all) and then of the 90 A-1s to the Luftwaffe took place in the middle of 1941, later than the RLM had expected but still earlier than the scheduled date. In fact, because of numerous accidents and the negative opinions of the pilots concerned, the order was given to suspend production completely at the beginning of 1942. The affair caused a scandal and added even more to the traditional enmity between

Me 210A-1 (VN AT) in flight.

Messerschmitt and Erhard Milch, recent successor to Udet as head of the RLM.

The halting of production did not prevent the continuation of the trials on the Me 210 and the V17 prototype which had the necessary changes to cure the defects: basically, automatic wing slots on the wing leading edges and increased depth of the rear fuselage. Messerschmitt re-proposed this new modified version and soon after, production began again with the modification of existing versions, the completion of numerous Me 210A-1s and A-2s, with production of the C version and finally with the Me 410, the definitive version.

The Me 210A-1, now modified, was joined by the A-2 (perfected for use as a dive-bomber), and became operational in the Channel in the Summer of 1942 and from the end of the year in the Mediterranean. The Me 210C was the version with 1475 HP DB 605 engines; it was built under license in Hungary (a third of Hungarian production was destined for the Luftwaffe). The Hungarian Air Force used the 'Ca-1' as a heavy fast bomber. It was very popular owing to its efficiency and optimum performance. 267 were produced and entered service at the beginning of '44.

Very similar to the Me 210 but possessing powerful 1750 HP DB 603A engines, the Me 410V-1 flew in the Autumn of '42, followed by numerous prototypes, while from the beginning of '43 supplies of the A-1, a fighter-bomber, and the A-2 'Zerstörer' began arriving. There were many sub-variants, including, in particular, the A-2/U4 armed with a BK 5 cannon of 50 mm for long-distance firing against bombers. It was used with a fair amount of success.

The A-3s were photographic reconnaissance planes. The Me 410B differed from the others in its 1900 HP DB 603G engines. The B-1 and B-2, respectively fighter-bomber and 'destroyer', were

Me 410 'Hornisse' at ground.

followed by the B-5 and B-6 antishipping versions, the latter being fitted with FuG 200 'Hohentwiel' search radar.

Over 1000 Me 410s were built and used profitably on all the fronts where the Luftwaffe was engaged. The Me 410 was also offered, in 1943, to the Italian 'Regia Aeronautica' but without success, while a Me 210 with DB 601 engines but brought up to 410 standards was supplied to Japan for trials.

Messerschmitt Me 210 A-1
Two-seat heavy-fighter and dive-bomber

Power plants: two Daimler-Benz DB601F 12-cylinder inverted-vee liquid-cooled engines, each rated at 1350 HP for take-off.
Dimensions: span 16.34 m; length 11.12 m; height 4.39 m; wing area 36.2 m^2.
Weights: empty equipped 7070 kg; max loaded 9700 kg.
Performance: maximum speed 465 km/h at sea-level, 563 km/h at 5400 m; service ceiling 8900 m; climb rate 9 m/sec; max. range 1820 km.

Armament: two 20 mm MG 151 cannon with 350 rpg; two 7,9 mm MG17 machine-guns with 1000 rpg; two 13 mm MG131 machine-guns in remotely-controlled barbettes, with 450 rpg (plus 1000 kg, max. internally carried bomb load).

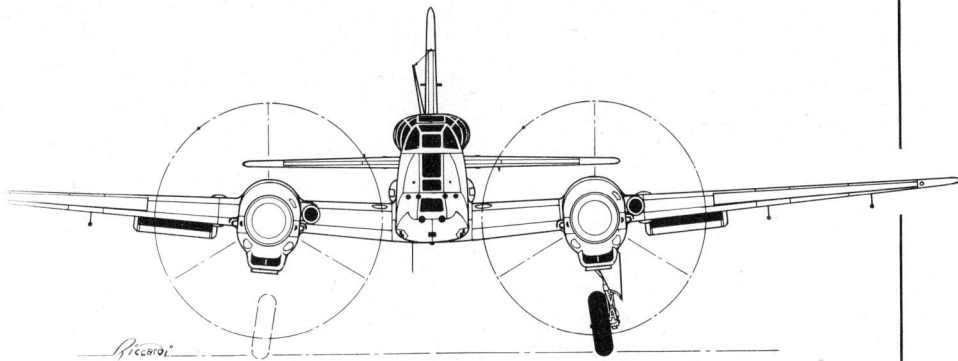

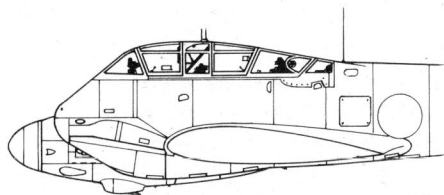

Me 410A-3

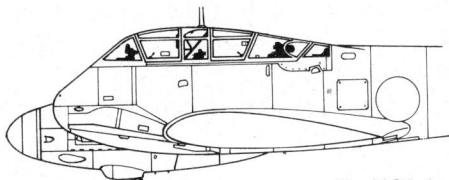

Me 410A-1

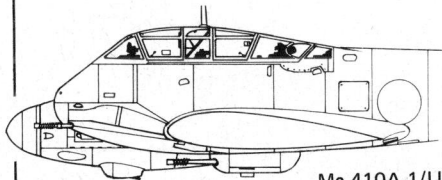

Me 410A-1/U2

Me 410A-1

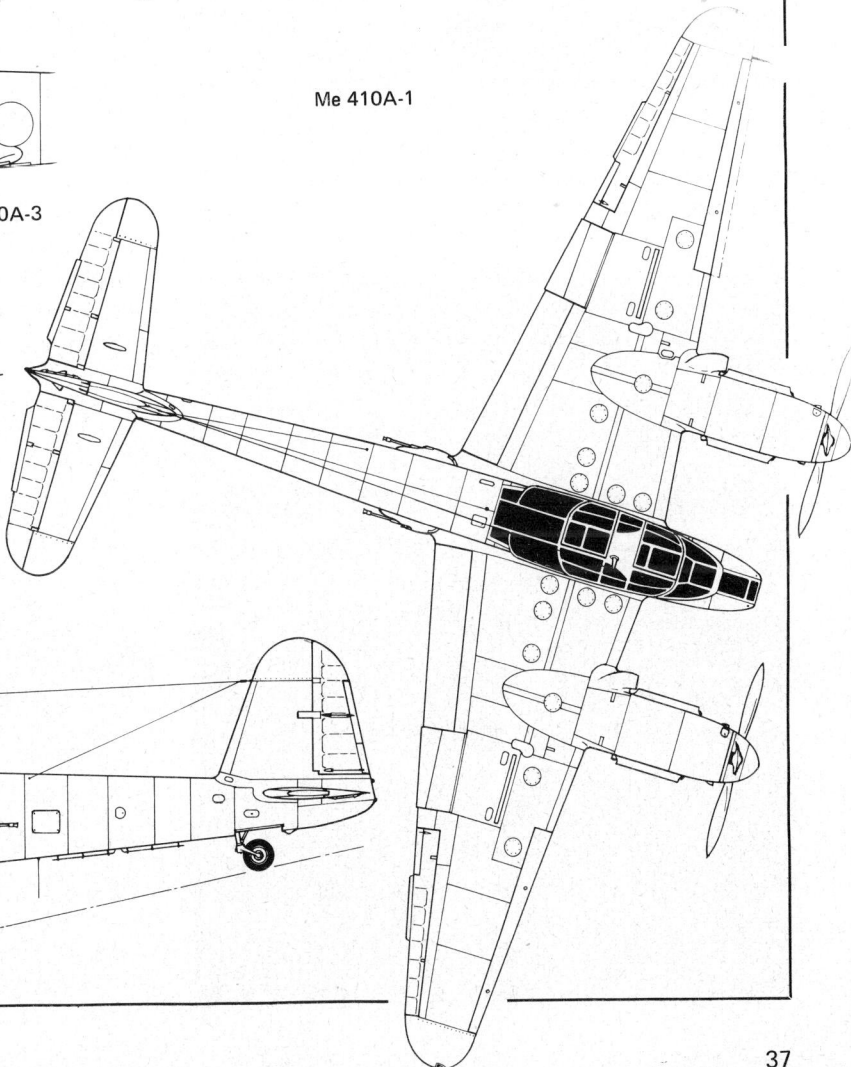

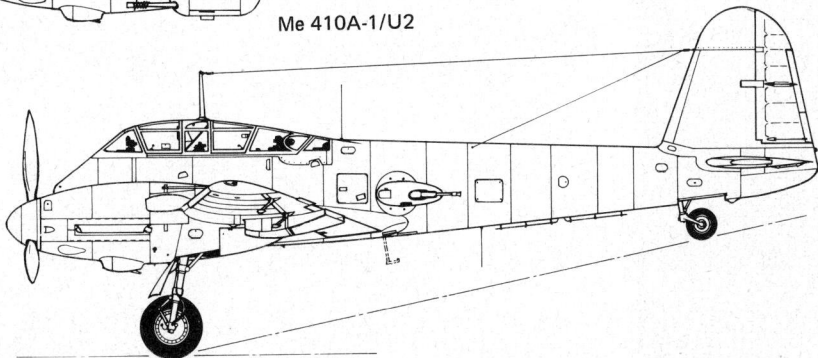

MESSERSCHMITT Me 163 'Komet'

From top to bottom: the first prototype Me 163V1 Komet (KE - SW) painted entirely in 'Hellblau 65'. One of the prototypes in a vertical climb.

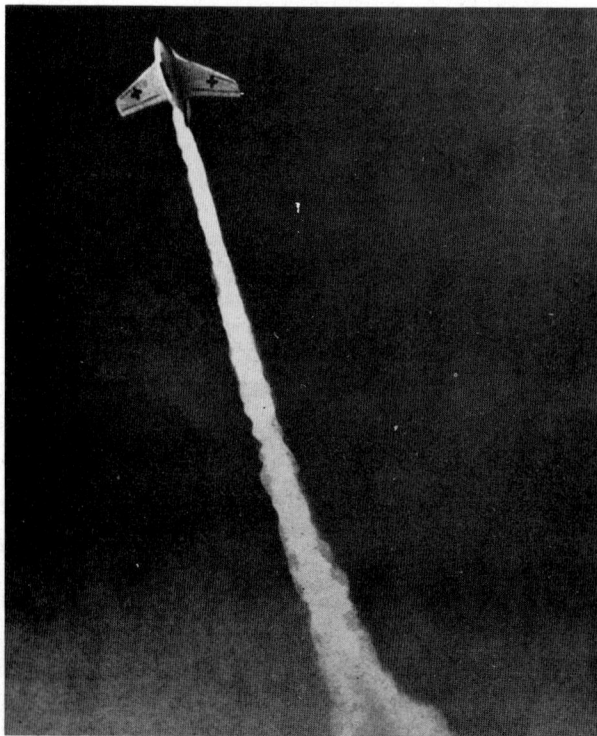

The Messerschmitt Me 163 Komet merits a special place in aviation history for being the first piloted plane to exceed 1000 km/h in horizontal flight and for being the only example of a combat aircraft with rocket propulsion in operational service. This sensational little plane originated from a design in the tailless aircraft category, developed by Dr. Lippisch under the aegis of DFS at the beginning of the 1930s.

Its development was then entrusted by the RLM to Messerschmitt, who grave it the designation Me 163. To achieve the extremely high speeds planned, Messerschmitt relied on the new liquid fueled rocket engines perfected by the Walter company of Kiel.

Designed in light of the valuable experience

Me. 163B V2

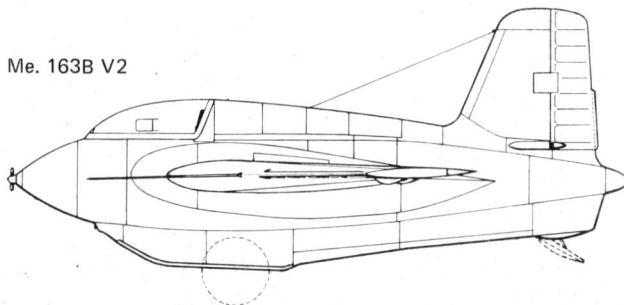

Me. 163B V6

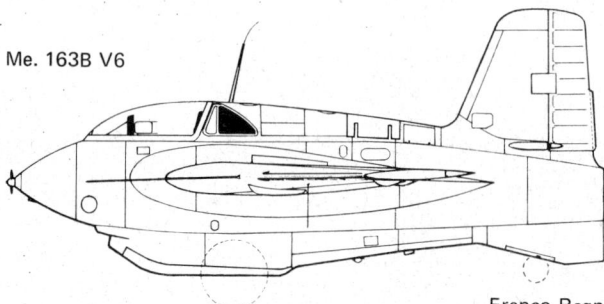

Franco Ragni

Above: the Me 163B-1a of the Deutsches Museum. Below:
a shot taken with a gun camera of an American fighter
and heavily re-touched (according to some a fake).

The Me 163B-1a used by the British RAF as 'VF241'.

gained in the field of low speed flight with the
fairly similar DFS 194, driven by a HWK-R-I-203
of 400 kg st (880 lb st), the first prototype
Me 163V1 made its appearance. It flew at first
with no engine, but in the Summer of 1941 the
HWK-R-II-203b of 750 kg st (1650 lb st) was
fitted, and in July the Me 163 began its life as an
experimental plane for high speed research.

The flights were entrusted to the famous glider
pilot Heini Dittmar, who on 2 October 1941
became the first man to exceed 1000 km/h (620
mph) in horizontal flight.

The Me 163 had a rather small, squat appearance
with a large mid swept wing. For take-off it used a
jettisonable dolly, and the landing was made using
a retractable ventral skid. Flight endurance was
only a few minutes due to the extremely high
consumption of fuel, a mixture of so-called 'T

Me 163V-41, Maj. Wolfang Spaete, Kommandoführer of Erprobungskommando 16, Bad Zwischenahn May 1944.

Me 163B-1a, 2/JG-400, Brandis, 1945

Me 163B-1a. Bad Zwischenahn.

Me 163B-1a., of an unidentified unit (13 or 14 Ergaenzungstaffel at Underfeld?)

Enzo Maio

Stoff' (hydrogen peroxide and water) and 'Z Stoff' (calcium permanganate and water).

As a result of the success of the prototype Me 163, a small series of similar models, designated Me 163A, was ordered. These were to serve basically as trainers for the interceptor version (Me 163B) designed around the projected 1700 kg st (3750 lb st) Walter HWK 109-509 engine, using as as a catalyst 'C Stoff' (hydrazine hydrate and methyl alcohol) instead of the 'Z Stoff', which promised greater working stability. Hopes of being able to bring the new fighter into service at the beginning of '43 were thwarted due to the long, difficult work of perfecting the new engine, which was not ready to be flight tested on one of the Me 163B prototypes until August of '43. The delay was such that the order for 70 pre-production examples was already partially completed without the engines being ready. About half the Me 163B pre-production models were designated 'Versuchs' (experimental) and the others were ordered as Me 163B-Os.

The appearance of the 'B' was different from that of the 'A'; the fuselage was bulkier, as it had to hold the operational equipment and a larger quantity of propellant; the wing was still built of wood and had a 23.3 deg. sweep angle at quarter-chord. After the first flights in the Spring of '42, fixed slots were adopted for a major part of the wing leading edge.

The Me 163B-O was armed with two small MG151 20 mm cannons, while the examples of the Me 163B-1a series were given the MK108 30 mm gun; otherwise the two models were almost exactly alike. Mass production was entrusted to Klemm Flugzeugwerke which organized a good network of subcontractors around a well-concealed assembly center.

The construction standards of the production models proved to be poor, unfortunately, a fact which together with the difficulties with the Walter engine, led to further delays in operational training of the first squadrons equipped with the Me 163.

In fact, for some time previously, EK 16 had been formed around a small élite cadre of pilots, to train them and explain the operating techniques for such a unique plane. Created in the style of the Peenemünde program, EK 16 was one of the first victims of the air offensive directed against 'secret army' bases by the Allies, following the revelations of photographic reconnaissance flights which kept German territory under constant surveillance. Bad Zwischenahn was chosen as a safer base for continuing these activities.

The experimental training program of EK16 (used with all subsequent units) consisted of first a period on Habicht (Hawk) gliders with progressive reduction of the wing span, aimed at familiarizing the pilot with the methods of accurate landings at high speeds (the Me 163 landed at 180 km/h-110 mph). Next they went onto an Me 163 with no engine and weighted with ballast before going on to the motorized version, preceded by an intense theoretical course of study. In May 1944 Staffel I/JG 400 was formed at Wittmundhaven, and given part of the experimental work in preparation for the production Komets, which were still not completely ready technically. Hurriedly, the other Staffeln of the Geschwader were formed, these scattered divisions being part of a plan designed to 'block the frontiers' to the invaders by locating the Me 163 formations in a series of bases fanned out over the territory.

The plan finally adopted was different; this plan concentrated all the Me 163s of EK 16 and JG 400 in Brandis near Leipzig for the defence of the synthetic fuel distilleries of Leuna.

During this period, the revolutionary Komet entered into combat. Its appearance among the American bombers caused great alarm because of its new type of propulsion and the Komet's sensational speed both horizontally and when climbing. In reality, the Me 163 pilots were only favored in theory as the difficulties they encountered operating their new plane were far greater than the apparently insuperable problems facing the American escort fighters.

In fact, once they had climbed to a higher altitude than the bombers, the Komets had an endurance of only 2 1/2 minutes which had to be managed with extreme prudence; the attack was usually made during an unpowered dive at a speed of approximately 1000 km/h (over 600 mph). At such a high speed they had no more than three seconds to fix their targets in the firing zone permitted by the heavy MK108; these weapons had enormous destructive power but too slow a rate of fire for such a brief time, and they frequently jammed. The HWK 109-509A, which was never satisfactorily perfected, was often subject to working defects including frequent engine cut-outs (very dangerous in take-off), while in landing (on a skid), the remaining 'C' and 'T' Stoff sometimes came into contact, resulting in an explosion and the destruction of the plane. Furthermore, after the first series of well-trained pilots, there arrived a mass of recently conscripted men, full of enthusiasm but insufficiently trained, so that the number of accidents rose considerably.

The efforts of the German technicians to improve the Komet were concentrated on the engine and armament.

For the latter, 24 R4M rocket missiles were

Messerschmitt Me 163B-1a Komet cutaway drawing key

1   Generator prop
2   Generator
3   FuG 25a radio
4   Cockpit air inlet
5   Reflector gunsight
6   Internal armored
    windshield
7   Cockpit canopy
8   C-Stoff tank
9   Fixed Handley-Page
    slat

10 Starboard wing
11 Wing box
12 Elevon (fabric-covered)
13 Fixed trim tab
14 Elevon control runs
15 Dive brake
16 Main C-Stoff wing tank
17 Flap
18 Control stick
19 Pilot's seat
20 Headrest
21 FuG 16ZY transceiver
    antenna
22 T-Stoff tank
23 T-Stoff filler
24 Port cannon ammo
25 Starboard cannon ammo
26 Filler

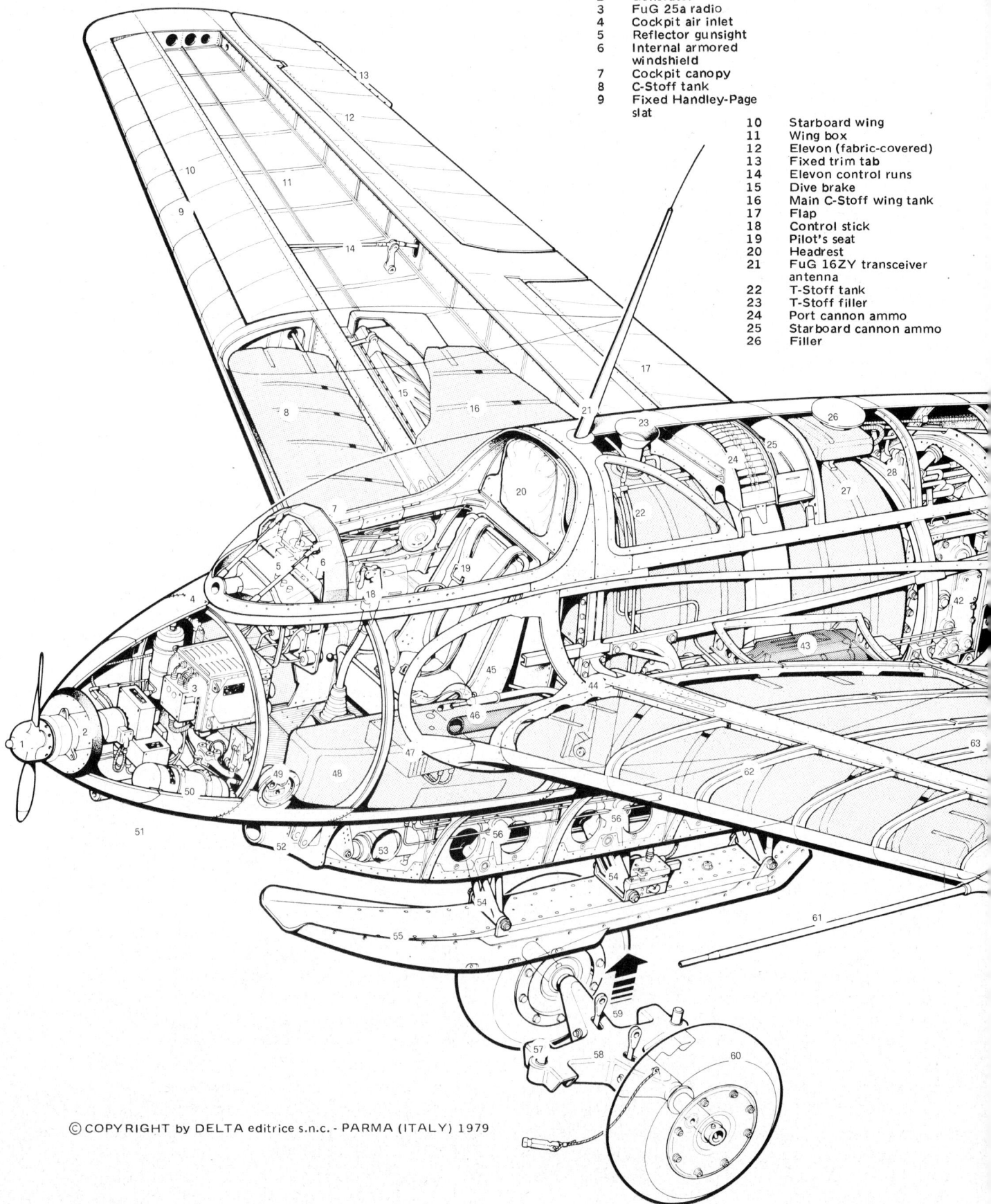

27 T-Stoff tank
28 Fuel valve
29 Fin construction
30 Rudder construction
31 Fixed trim tab

32 Rudder control run fairing
33 Combustion chamber
34 Exhaust duct
35 Exhaust nozzle
36 Fuel vent
37 Shock absorber
38 Tailwheel retraction actuator
39 Floating structure
40 Tailwheel fork

41 Engine mount
42 Injector control system
43 MK108 cannon breech
44 Wing/fuselage bolt
45 Wing/fuselage fairing
46 Cannon muzzle
47 FuG 25a IFF
48 T-Stoff tank

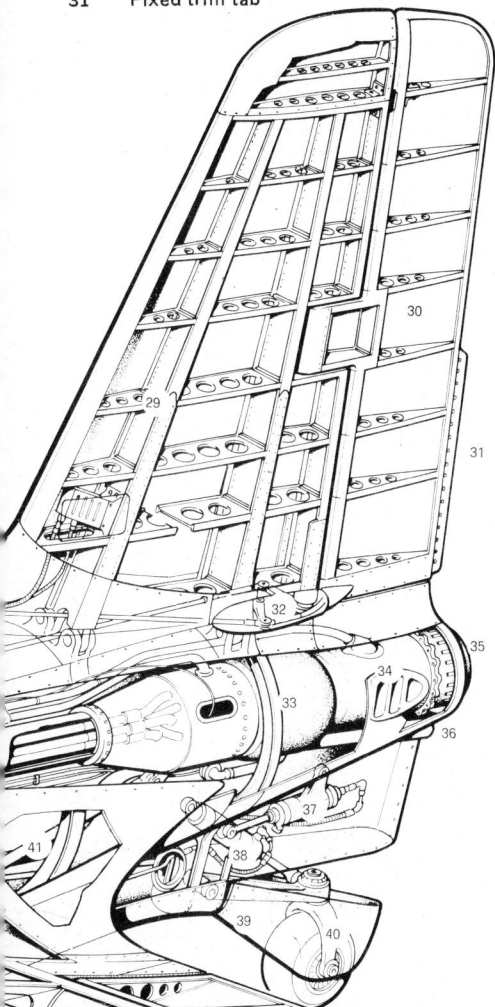

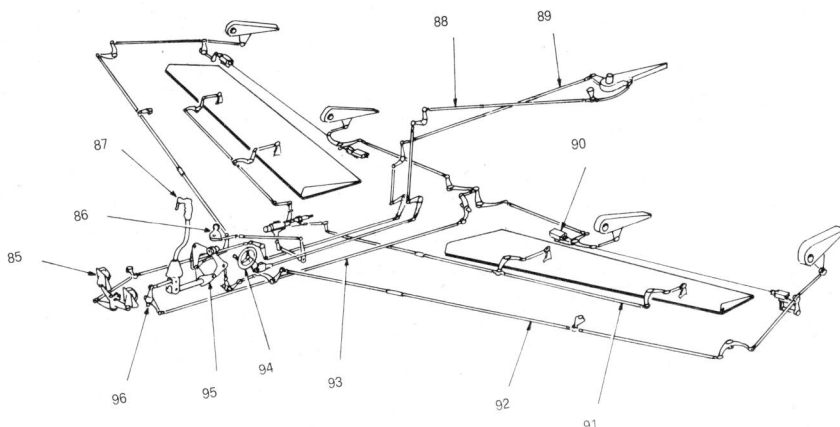

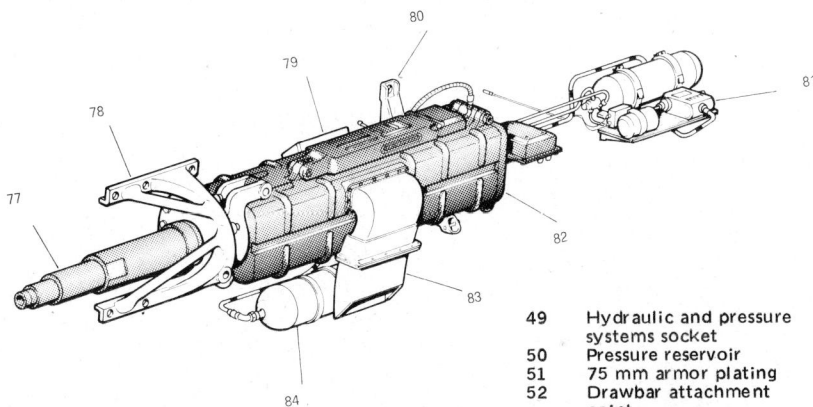

49 Hydraulic and pressure systems socket
50 Pressure reservoir
51 75 mm armor plating
52 Drawbar attachment point
53 Landing skid actuator
54 Skid brackets
55 Retracting skid
56 Skid structure
57 Dolly line-up pinion (for take-off)
58 Dolly
59 Dolly retainer
60 Wheel
61 Pitot tube
62 C-Stoff tank

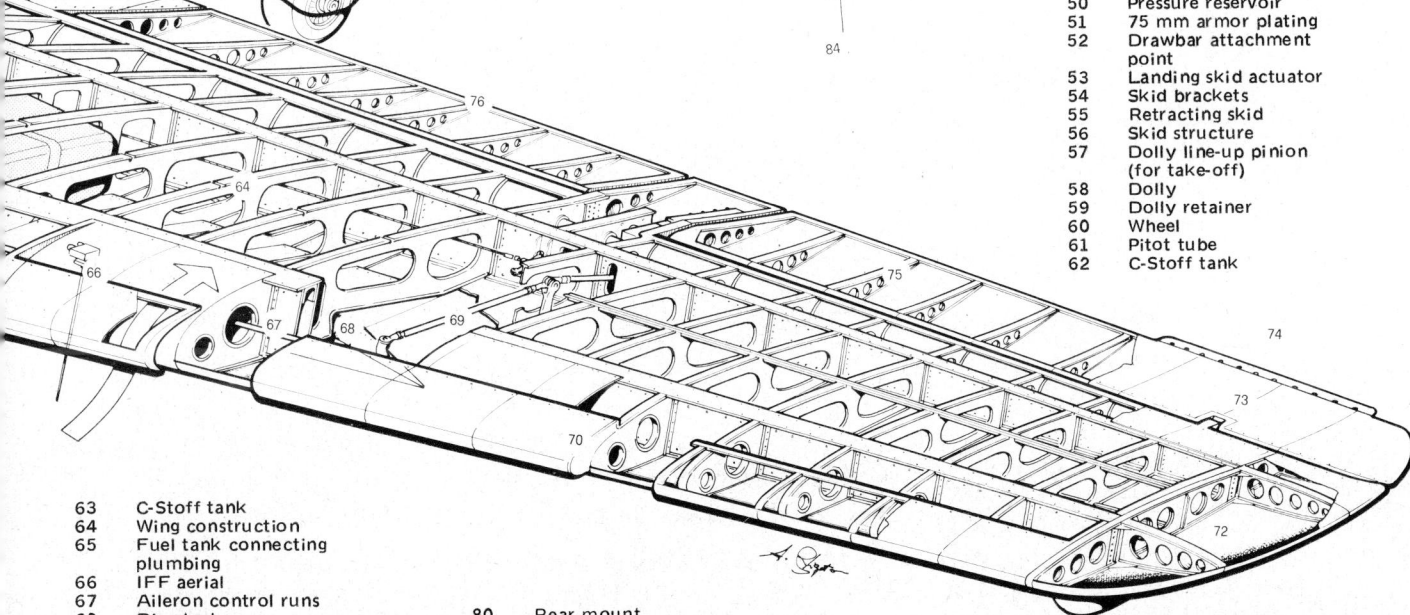

63 C-Stoff tank
64 Wing construction
65 Fuel tank connecting plumbing
66 IFF aerial
67 Aileron control runs
68 Dive brake
69 Aileron control runs
70 Fixed slat
71 Landing skid
72 Wingtip
73 Elevon
74 Fixed trim tab
75 Elevon construction
76 Flap construction
77 Cannon muzzle
78 Front mount
79 Ammo feeding

80 Rear mount
81 Cocking and firing mechanism
82 30mm MK108 cannon (port)
83 Spent case chute
84 Pneumatic re-cocking bottle
85 Rudder pedals
86 Dive brake control lever
87 Control stick
88 Rudder lock control runs
89 Rudder control runs

90 Differential elevon movement lock
91 Dive brakes control runs
92 Elevon control runs
93 Flap control runs
94 Rudder lock control wheel
95 Aileron control trapeze
96 Joint

43

Me 163B-1a

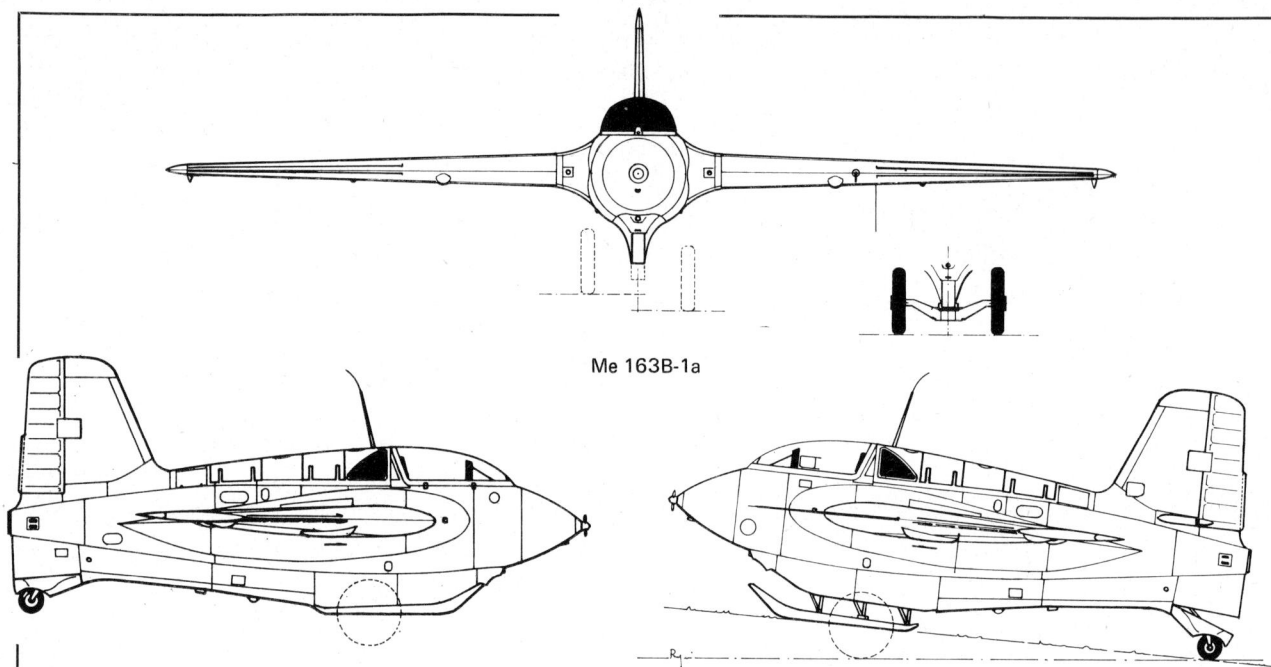

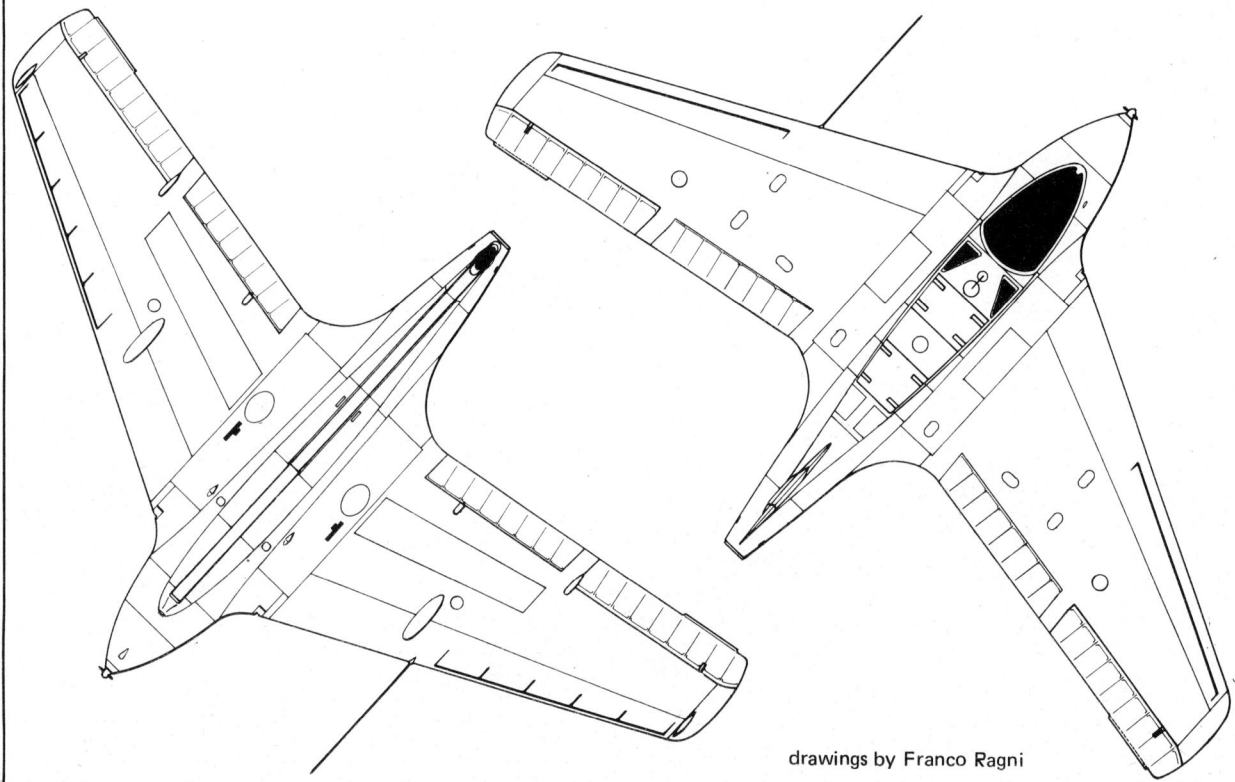

drawings by Franco Ragni

Messerschmitt Me 163B-1a Komet

Day point interceptor, single-seat

Power plant: one Helmuth Walter HWK-509A-1 or A-2 liquid fuel rocket rated at 1,700 kg st. Fuel capacity: 1,527 liters.

Dimensions: wing span: 9.32 m; length: 5.70 m; height: 2.75 m; wing area: 19.60 sq m.

Weights: empty, equipped: 1,900 kg; max. loaded: 4,300 kg

Performance: max. speed: 960 km/h at 9,000 m (Mach 0.88 approx.), 835 km/h at sea level (Mach 0.68);

mission speed (between 3,000 and 9,000 m, average): 796 km/h; initial climb rate: 81 m/sec; service ceiling: 12,000 m; absolute ceiling: 15,100 m; climb to 9,000 m: 2 min 36 sec, to 12,000 m: 3 min 21 sec; max. range: 33-100 km; endurance: 2.5-7.5 min.

Armament: two Rheinmetall-Borsig 30 mm MK108 cannons forward firing, in the wing roots, with 60 r.p.g. Twelve Me 163B-1a also had 10 'Jagdfaust' unguided air-to-air rockets.

tested, but of greater interest was an experimental armament with 10 rockets buried in the wing structure for oblique upward firing. As for the engine, an improved model, the HWK 109-509C was developed; it differed from the 'A' by its extra combustion chamber or 'cruising chamber', capable of supplying 300 kg st (660 lb st) independently, or with pressure from the main chamber. In this way endurance was increased by a few minutes. The new engine was meant for the Me 163C and Me 263, improved versions of the Komet. The latter had a normal retractable tricycle landing gear.

The end of the War came before the solution of the Komet's problems, however, and a high price was paid for the few victories won. In all, little more than 350 models (of this plane) were built, but its history did not end with the German surrender; in fact, the Japanese bought the production rights of the Me 163, building the J8M1 for the Marines and Ki-200 for the army. The only one to fly was a J8M1 on 7 July 1945, which unfortunately crashed immediately after its first take-off. After the War, Russia built a version of the Me 263 with straight wings and a conventional tail assembly as the Mikoyan I-270; there was no follow up, however.

One little known particular: in January of '45, a group of Italian pilots from the RSI (Repubblica Sociale Italiana) received preliminary training on the Habicht glider, and theoretical training on the Komet, but the deterioration of the military situation prevented completion of the training program and the introduction of the Me 163 into operational service with the RSI.

MESSERSCHMITT Me 262

The twin-jet Messerschmitt Me 262 was certainly the most advanced of all the fighters used in the Second World War, and appeared on the bloody European battle scene as a preview of the future of aviation. It was denied the privilege of being the first jet fighter conceived as such by the Heinkel 280, but the Me 262 was the first to achieve operational status.

Like all innovative planes, it did not get off the ground easily, but to be frank, who could blame those who would not give priority to the new type of propulsion when it seemed that the War would be won easily with conventional, excellently conceived planes like the Bf 109 and the Fw 190? No one could read the future, and when this became a preoccupation, the decisions taken were not the finest examples of lucid thinking. The excellent organizational efforts of German industry were wasted because of the inadequacy of the instructions at top political levels. The concept-development and production effort was, however, exceptional, and perhaps no other nation would have succeeded better than the Germans, given the same conditions.

In the years immediately preceding the War, the design studies of jet propulsion in Germany were already at a good point (as they were in other countries, too) following independent project lines: centrifugal turbojets at Heinkel (the first turbojet propelled plane was the He 178, flown 27 August 1939), and axial-flow turbojets at BMW and Junkers.

The RLM had the latter type in mind when it asked Messerschmitt to design and produce the jet fighter which eventually became the Me 262. In its final form the design projected an elegant twin-jet with a triangular-section fuselage, underslung engines on a swept wing. The landing gear was the conventional type with a tail wheel. Three prototypes were ordered. The first, the Me 262V1 flew on 14 April 1941 with a 700 HP Jumo 210G piston engine instead of the BMW 003 jets which were well behind schedule in their development. The Jumo 004 was in an even less-advanced state of development.

In this condition, the Me 262V1 made only seven flights in seven months to gain some experience in the low speed flight envelope and when, at the end of the year, the long awaited 003 jet engines were mounted, Messerschmitt decided they preferred to keep the Jumo 210 in its place. This unusual 'trimotor' flew for the first time on 25 March 1942, and was able to land thanks to its piston engine alone as both jet engines had failed. While BMW was re-designing the 003 completely, the bigger and more powerful Jumo 004A became available, and fitted with two of these 830 kg-st (1830 lb-st) engines, the Me 262V-2 flew on 18 July 1942, this time with success. Meanwhile the '262V1 continued to fly with the old piston engine.

Other prototypes had been ordered in the meantime, with various aerodynamic modifications suggested by earlier test results; the

tricycle undercarriage made its appearance on the Me 262V5 and was adopted as standard.

In the Luftwaffe staff, the first to realize the possibilities offered by the new plane was the 'General der Jagdflieger', Adolf Galland, but although he managed to have the number of pre-production models increased from 15 to 30, no efficient or coherent decision followed regarding the mass production versions, due to political interference, with directives, coming one moment from Milch, then from Goering, and even Hitler himself. But even the most favorable decision would have been premature, given the difficulties experienced by Messerschmitt in organizing mass production of the new design, and by Junkers in rendering more reliable the delicate Jumo 004 (now being tested as the 'B' version of 900 kg st-1980 lb st) more.

Above and below: two photos of the Messerschmitt Me 262A-1a Schwalbe, single seat day fighter version.

An example of these difficulties can be shown in the fact that the majority of the pre-production models were not used till the Spring of '44 because of the unavailibility of engines. In the meantime, Messerschmitt began preparing for the production levels asked for at the end of 1943, drawing up production data instructions and delivery dates which proved to be completely unrealistic.

After the 30 pre-production Me 262A-O models, 7 of which were designated 'Versuchs' (the last, the V12, was a special model for high speed flying, and had reached a speed of 1004 km/h-624 mph) in the Summer of '44, the first Me 262A-1a 'Schwalbe' (Swallow) appeared. This

The bomber version Me 262A-2a 'Sturmvogel' with two 550 lb bombs.

interceptor version had a 'troubled infancy', not so much because of the necessarily long development period, but rather because of an order from Hitler to give top priority to the new plane as a 'Blitzbomber'. In fact, the Führer was going through a period in which he was fixed on the idea of a reprisal against the Allies. The order was revoked in November but clearly several months had been lost, even if the bomber version, the Me 262A-2a 'Sturmvogel' (Stormbird), performed well on the whole. The new model differed from its predecessor (though practically they were contemporaries) in addition of two pylons under the forward fuselage for 2x250 kg (550 lb) bombs, and by the bomb sight in the cockpit. The 4x30 mm cannon armament was still fitted in the nose. The new plane was brought into service in the Summer of '44, as a fighter, by EK 262 (a service test unit formed in April to define suitable operational tactics, and for pilot training). Soon after, in August, the bomber version issued to the KG 51 made its debut. In October, despite being prohibited to use the Me 262 as a fighter, EK 262 reformed as an independent unit known as Kommando Nowotny after the famous ace who commanded it, and engaged in a series of actions against American bomber formations.

If the results were not exceptional, they were not disappointing either, and a month after coming into service, on the death of Nowotny, the surviving pilots and planes reformed as a Gruppe of JG7, a fighter unit which used the Schwalbe more aggressively. Other groups were formed to complete the Geschwader, and other units were

added in the last months of the War; even the Sturmvogel bombers increased in number. In February of 1945 JV 44, the most famous unit formed around the Me 262, began training, commanded by Galland himself, who had under his command the élite of the Surviving German fighter aces.

Following the initial experiences in combat the new Me 262 A-1b was introduced to these units, featuring an integrated armament of 24 50 mm R4M rockets hung from two wing racks. The rockets made it possible to hit targets from a great distance, thus increasing the range and time available to destroy enemy aircraft, and reducing the danger from enemy defensive fire.

In frequent encounters with Allied fighters, the latter made several 'kills' due to the most diverse circumstances but above all because the Allies had learned to attack the German jet planes during landings when they were most vulnerable.

In all, about 1400 Me 262s were built, but of these only a small percentage became operational. Their high performance gave excellent results, not with standing the innumerable problems involved in bringing such a revolutionary plane into service, problems which were only partially resolved.

Production standards, especially with the engines (excellent in performance and economical, but with an extremely short service life), were never very high, given the critical situation of German industry in the last phases of the War.

Me 262 A-1a Wk Nr. 111 857

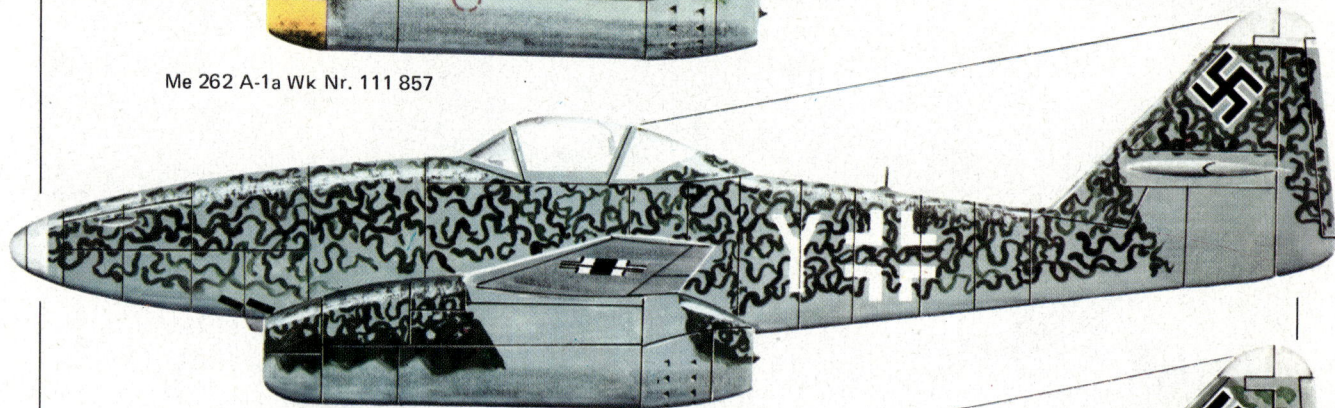

Me 262 A-2a «White Y» I/KG 51 - Achmer 1945

Me 262 A-2a I/KG 51

Me 262 B-1a/U1 10/NJG 11 - Leck-Holstein

Me 262 A-1b 9/JG7 – Dübendorf 25 April 1945

S. Jamois

48

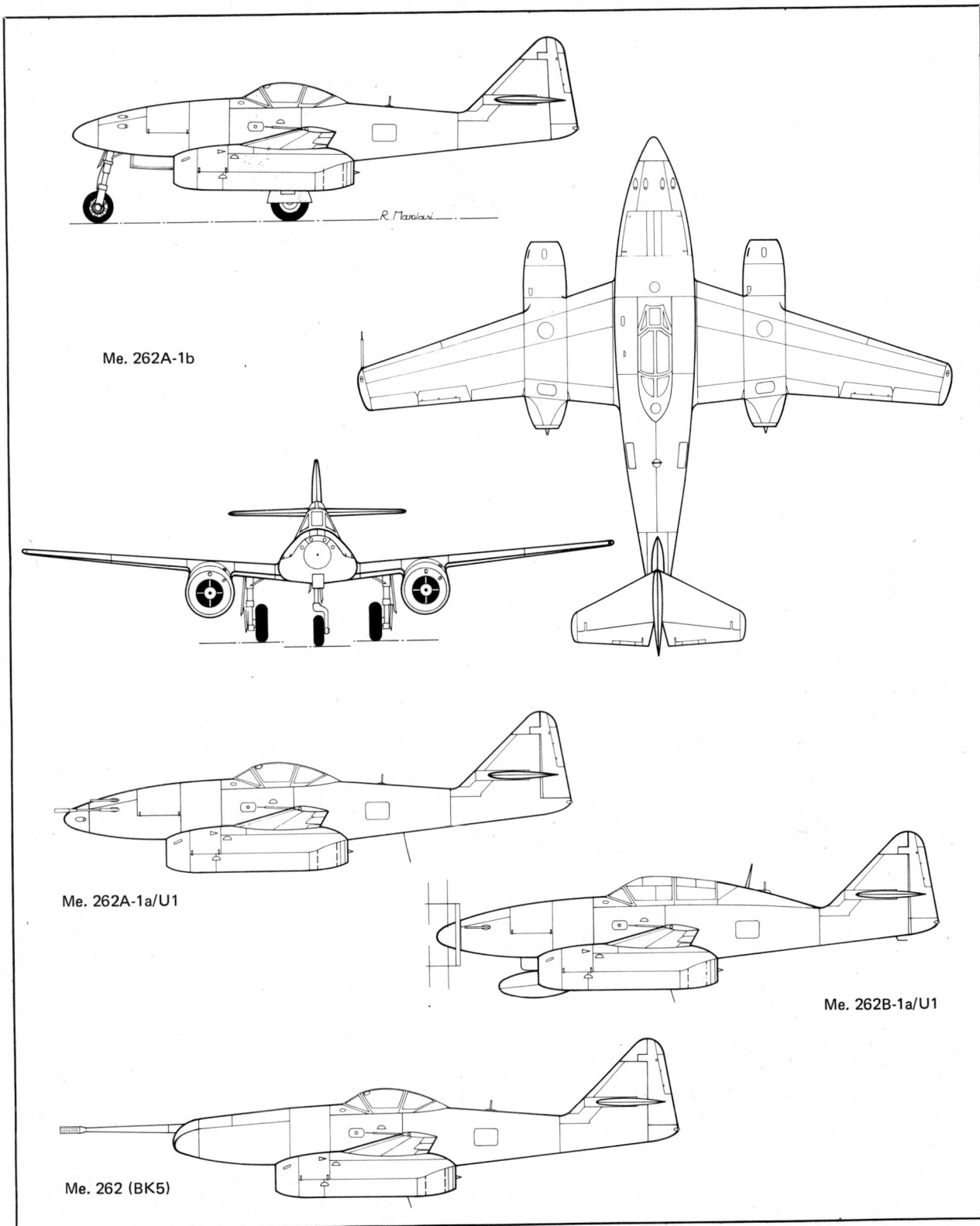

Me. 262A-1b

Me. 262A-1a/U1

Me. 262B-1a/U1

Me. 262 (BK5)

Many of the pilots, too, left something to be desired, being taken from the last conscripts and insufficiently trained, but also because of the mistaken belief that the Me 262 was a relatively 'easy' plane and did not require specific training. In fact a tandem two-seat training version, designated the Me 262B-1a appeared only at the end of '44. From this a night fighter variant was developed and a few models brought into service in the last days of the War with FuG 218 'Neptun' radar. The final version of the Me 262, the B-2a night fighter, was too late to enter mass production. It could be identified by the longer fuselage, which contained new fuel tanks to replace those eliminated in the Me 262B-1a to make room for the second crewman.

The last of the major versions was the 'C' which remained experimental and demonstrated a vast

Messerschmitt Me 262-A Schwalbe cutaway drawing key

1. Gun camera port
2. Gun camera
3. Nosewheel door
4. Retracting jack
5. Shock absorber retainer
6. Nosewheel leg
7. Nosewheel
8. Cannon muzzles
9. Hydraulic plumbing
10. Four 30 mm cannons
11. Spent case chute
12. Bulkhead
13. Compressed air bottles
14. Four 30 mm cannons
15. 900 liter armored tank
16. Fuel filler
17. Engine cowling
18. Air intake
19. Rudder pedals
20. 170 liter tank
21. Wheel housing
22. Cockpit skin
23. Pilots seat
24. Armored headrest (15 mm)
25. Starboard-hinged canopy
26. Armored windshield (90 mm)
27. Revi 16B reflector gunsight

28. Instrument panel
29. 900 liter armored fuel tank
30. 600 liter fuel tank
31. FuG 16ZY radio
32. Master compass
33. Control runs
34. Fuel filler
35. FuG 25a IFF aerial
36. Rudder control runs
37. Tailplane adjustment
38. Trim tab gear
39. Tailplane construction
40. Rudder
41. Flettner tab
42. Navigation light
43. Elevator
44. Flettner trim tab
45. Navigation light
46. Frise aileron
47. Flap
48. Automatic slats

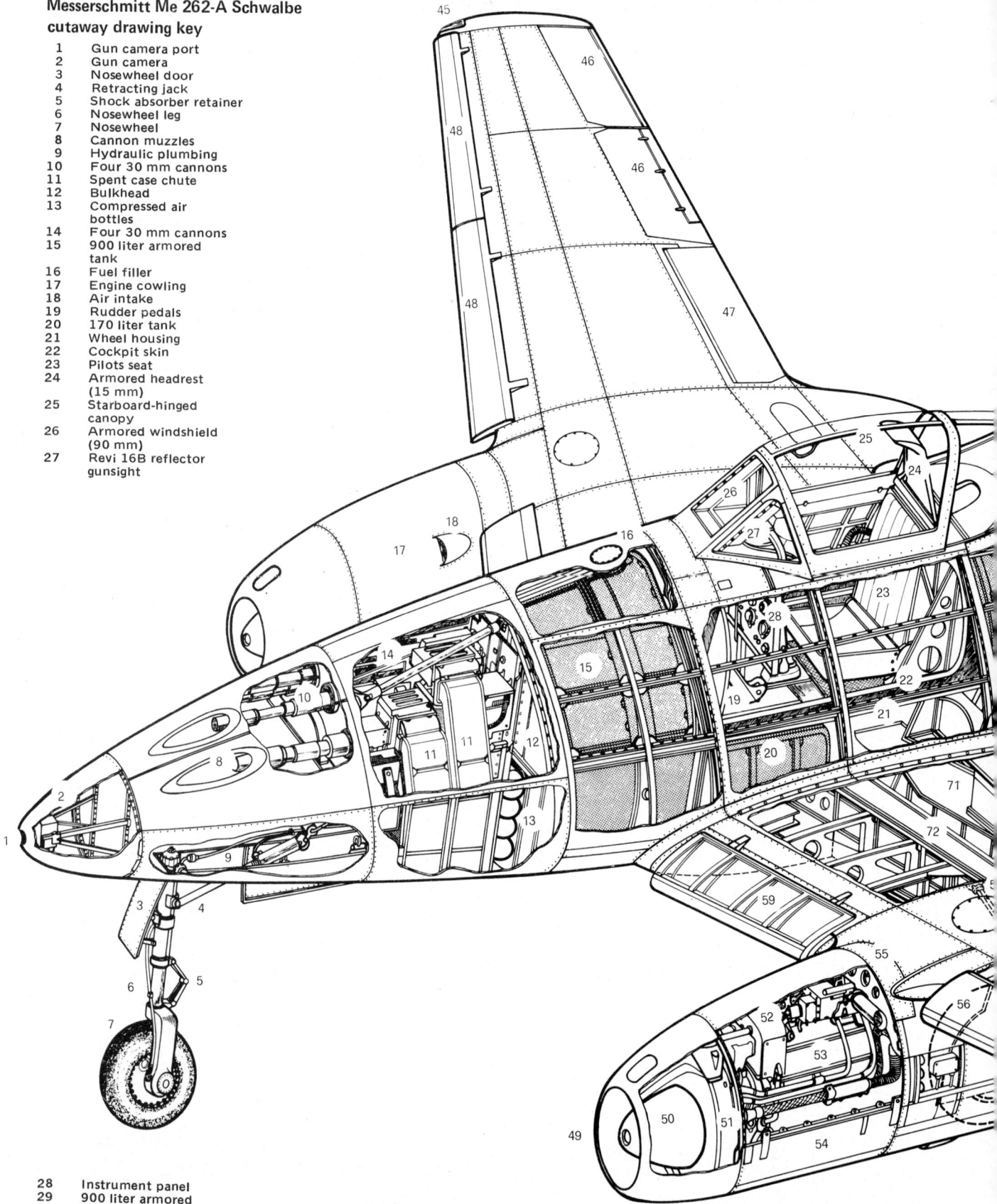

49 Air intake
50 Riedel starter
51 Oil tank
52 Accessories carter
53 Jumo 004B-2 turbojet
54 Opening cowling
55 Engine mount
56 Mainwheel
57 Actuating rod
58 Undercarriage fairing
59 Slat construction
60 Engine mount wing rib
61 Flap construction
62 Exhaust nozzle
63 Flap
64 Frise aileron

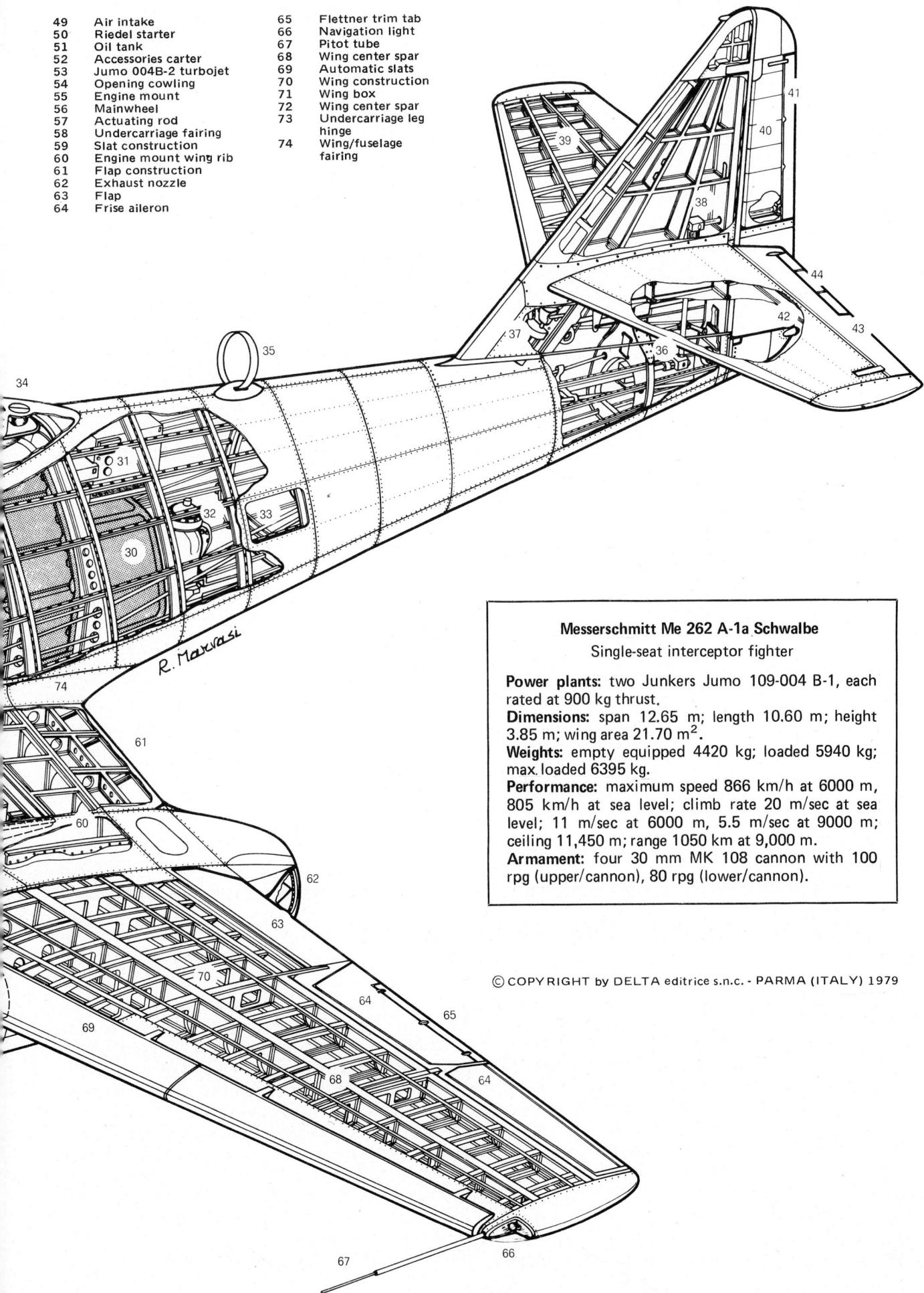

65 Flettner trim tab
66 Navigation light
67 Pitot tube
68 Wing center spar
69 Automatic slats
70 Wing construction
71 Wing box
72 Wing center spar
73 Undercarriage leg hinge
74 Wing/fuselage fairing

R. Maxvasi

Messerschmitt Me 262 A-1a Schwalbe
Single-seat interceptor fighter

Power plants: two Junkers Jumo 109-004 B-1, each rated at 900 kg thrust.
Dimensions: span 12.65 m; length 10.60 m; height 3.85 m; wing area 21.70 m^2.
Weights: empty equipped 4420 kg; loaded 5940 kg; max. loaded 6395 kg.
Performance: maximum speed 866 km/h at 6000 m, 805 km/h at sea level; climb rate 20 m/sec at sea level; 11 m/sec at 6000 m, 5.5 m/sec at 9000 m; ceiling 11,450 m; range 1050 km at 9,000 m.
Armament: four 30 mm MK 108 cannon with 100 rpg (upper/cannon), 80 rpg (lower/cannon).

© COPYRIGHT by DELTA editrice s.n.c. - PARMA (ITALY) 1979

A sample of the twin seat night fighter captured by the RAF.

increase in climb rate thanks to the adoption of a single rocket motor such as the HWK 109-509 (mounted in the tail or mounted externally and being jettisonable), or the BMW 718 rocket motor which, with the BMW 003A turbojet, comprised the BMW 003R mixed power plant, which in this case replaced the Jumo 004 in the nacelles.

Auxiliary rockets had already appeared on the Me 262 production models to assist in take-offs. The units used were Borsigs, of 500 kg st (1100 lb st).

There were a huge number of experimental Me 262s, fitted with various types of armament and equipment. Experiments concerned 50 mm cannon and 30 mm cannon in the 'Schräge-Musik' position, equipment for photgraphic reconnaissance, bomb sights for a bomb ardier positioned in a glazed nose, etc. In Czechoslovakia after the War, the Avia firm undertook limited production of the Me 262, re-numbered S.92, using equipment and components already built in Czechoslovakian factories during the German occupation.

At the end of the War, personnel of the 2nd Gruppo Caccia (fighter group) of the RSI (Repubblica Sociale Italiana) were on the point of being transferred to Germany for training on the Me 262.

FOCKE-WULF Fw 190 series

Virtually all the national air forces militarily involved in the last world conflict had at least two types of fighter aircraft in service at the beginning of hostilities: one usually more advanced technically, and the other a more traditional type as a safeguard against partial failure of the former. In addition, they generally possessed a fair number of planes from the previous generation of designs (in general biplanes with fixed undercarriage and mixed construction).

The sole exception was Germany, which arrived on the scene of battle with only the Messerschmitt Bf 109, probably the best fighter plane in the world, in normal front line service. However, the prototype Focke-Wulf Fw 190 was already flying, and was destined to support the 109 in the fighter units of the Luftwaffe. The Fw 190 is considered by many to be the most formidable fighter of the Second World War, considering its widespread use, the quantity built, and its performance. Its exceptional development potential helped it to keep abreast of the more and more complex demands which the War imposed on a fighter plane, thus remaining at the forefront of technical progress.

The Focke-Wulf Flugzeugbau, whose technical

director was the engineer Kurt Tank, was, in 1937, entrusted with the design of a new fighter for the Luftwaffe. Conscious of the Ministery's preference for in-line liquid-cooled engines, Tank submitted several designs with DB 601 engines, but succeeded in gaining the RLM's confidence with an alternative plane using a '1550 HP BMW 139 radial engine'. Three prototypes were ordered, the first of which flew on 1 June 1939. The Fw 190V1 was a compact and aggressive-looking plane. The undercarriage had a noticeably wide track (almost as if to underline the difference from the principal defect of the BF 109), but above all, it differed from the 109 by the adoption of a ducted spinner for the radial engine which faired over the front of the nose. Its behavior in flight was considered excellent by the test pilots, but the BMW engine gave a little trouble because of insufficient cooling. On the first and second prototypes, various solutions were tried out to avoid overheating of the engine and... the pilot, who sat immediately behind the firewall bulkhead. The final cowling configuration was a combination of a normal NACA cowling with a cooling fan on the reduction gear of the propeller. The Fw 190V2 was given two 7.9 mm guns, also.

In the meantime, BMW had developed, from the 139, the new 801, which although keeping the same frontal dimensions, was more powerful and demonstrated a far superior development potential. In addition, being longer and heavier it permitted the pilots to sit further back to solve a

c.g. problem, thus solving the problem of overheating in the cockpit. Moving the cockpit then permitted the adoption of synchronized weapons in the cowling. The V3 and V4 were cancelled, and the first prototype with the BMW 801 was the V5, built and tested in two forms: the V5k (klein-small) and the V5g (gross-large), with, respectively, normal wing spans and larger wing spans. The latter form was the one chosen as it had more pleasant handling characteristics.

The V6 and V7 were followed by a limited number of Fw 190A-0s, the first seven of which still had the reduced wing span. Many of the A-0s were given 'Versuchs' numbers, and the others, armed with MG17 7.9 mm guns, were used in operational evaluation tests.

In the Spring of '41 the delivery of 100 of the initial A-1 series was completed. The Luftwaffe received the first consignment in May and the first unit to be equipped with them was II Gruppe of JG26, based in Belgium. The 'baptism of fire' came in the Autumn of the same year with incredible results. The English found that their fighters were far out-classed by the new enemy design. Naturally the Fw 190 also had problems still to resolve, so much so that the first real production version was the Fw190A-2, which remedied the problem of insufficient armament found in its predecessor. The A-2 was armed with

two MG17 7.9 mm guns in the fuselage, two '20 mm MG151 cannon' in the wing roots and, on many models, two MG/FF 20 mm cannons in the wings. The engine was a BMW 801C-2 of 1600 HP.

While Great Britain was still delivering the early MkV Spitfire (awaiting the availability of more modern fighters) Focke-Wulf was continually turning out new models of their brilliant Fw 190 fighter from their factories in a production program that was going so well that in 1942 it already accounted for more than half the total monthly single-engined fighter output in Germany.

The A-3 version was produced, with the more powerful 1700 HP BMW 801D-2 engine, numbering more than 500 examples divided into numerous sub-variants, including especially the Jagdbomber (fighter-bomber) with various combinations of offensive loads. These sometimes resulted in the elimination of the two MG/FFs in the wings.

Up to October of 1942, the Fw 190s were confined to the more exacting Western front, given the inferiority of the Russian planes used against the Luftwaffe on the Eastern front. In that year, the victories of the Focke-Wulf were exceptional, and when the Jabo Fw 190A-3/U1 began its daily attacks on England, the RAF used

Left: a Fw 190A-1 pre-production model. Above: the Fw190V2 Werk Nr 0002.

its recently built Typhoons for chasing the fast-flying invader but to little effect. The typhoon's short nose often made identification difficult because of its similarity to the Fw 190.

The following model, the Fw 190A-4, differed from the others only in its methanol-water injection system, MW 50, which for brief periods permitted power output of about 2000 HP with the BMW 801, by injecting water and methyl alcohol. The sub-variants of the A-4 were primarily fighter-bombers while the Fw 190A4/R6 was armed with two 21 cm. mortars under the wings. Planes of this type were used with some success for long distance firing on American bomber formations to break them up. This left the bombers isolated and easy prey for the German fighters. In the meantime, the Fw 190A-4 had been introduced on the Russian front where it was extremely effective against the bombers and assault planes of the Russian air force.

The Fw 190 was also used on another front, but

The Focke-Wulf Fw 190 V5k (Kleiner Fluegel) with reduced wing span.

4 Fw 190A-O/U2 on the airfield at Bremen.

only for a short period. This was the African theater where it managed to take part in the operations in Tunisia.

The '190s then remained to hinder the Allied landings in Sicily and Salerno. At that time a few Fw 190s had already fallen into the hands of the Americans and British, and evaluation was in progress, resulting in a great deal of interest in the advanced features of the plane, and furnishing useful information and data for the development of new types of high performance fighter planes. For example, in England, the Hawker Fury project was influenced by the design of the Fw 190As tested during the war.

Variants of the Fw 190 continued up to the A-8 with numerous sub-variants which, conforming to German practice, were often available in pre-fabricated 'kits', so as to allow modifications of the basic type directly in the field. The engine remained the same, the BMW 801 D-2 with the MW 50 injection installation, which gave emergency speeds of up to 670 km/h (416 mph) with a 'clean' airplane. External modifications, made possible by the robust and adaptable airframe of the Fw 190, concerned above all more and more powerful armament with different combinations of 20 and 30 mm cannon, and continually heavier offensive loads, up to 1000 kg (2200 lb) as on the Fw 190 A-5/U-3.

The A-5/U2 was a night fighter version, following the operational tactics used in the Summer of 1943 and named 'Wilde Sau' (wild boar). The Messerschmitt Me 109 was also adapted for this job. However, the victories gained were very costly in terms lost planes and pilots. After a few months the Luftwaffe returned exclusively to twin-engined night fighters fitted with radar.

The Fw 190A-9, which did not go beyond the prototype stage, (V34) was equipped with armored wing leading edges to enable it to 'ram' the wings of enemy aircraft.

The numbering of Fw 190 variants was continued on the Fw 190D, starting with 'D-9'. The F and G versions were very closely connected with the Fw 190A. In fact, both merely appropriated as standard the modifications introduced on the Fw 190A to adapt it for use as a fighter bomber. The Junkers Ju 87 had changed from the rôle of Sturzkampfflugzeug (dive-bomber) to that of Schlachtflugzeug (close-support plane), but even in this new capacity it proved to be excessively slow and vulnerable and, above all, incapable of defending itself.

The Focke-Wulf Fw 190 had already proved itself to be an optimum aircraft for that particular rôle and from 1942 on, the Fw 190F and G were produced alongside the A, to substitute for the Stukas. A characteristic which differentiated many of the Fw 190Fs from the other versions with the BMW 801 engine was the blown cockpit

Fw 190A-3, 5./JG-1. Channel area April 1942.

Fw 190A-5/U15. Experimental unit for trials of LT-950 torpedo, 1942-43; the dotted line represents the BT-1400 experimental bomb.

Fw 190F-8/R3, SG2, Hungary, January 1945.

Fw 190A-8 captured by 414th Fighter Squadron, USAAF, St. Troud, Belgium, January 1st, 1945.

Fw 190A-8/R7 'Rammjaeger', 1./JG26. Kommandeur Major Karl Borris, Belgium, September 1944.

S. Lora Lamia

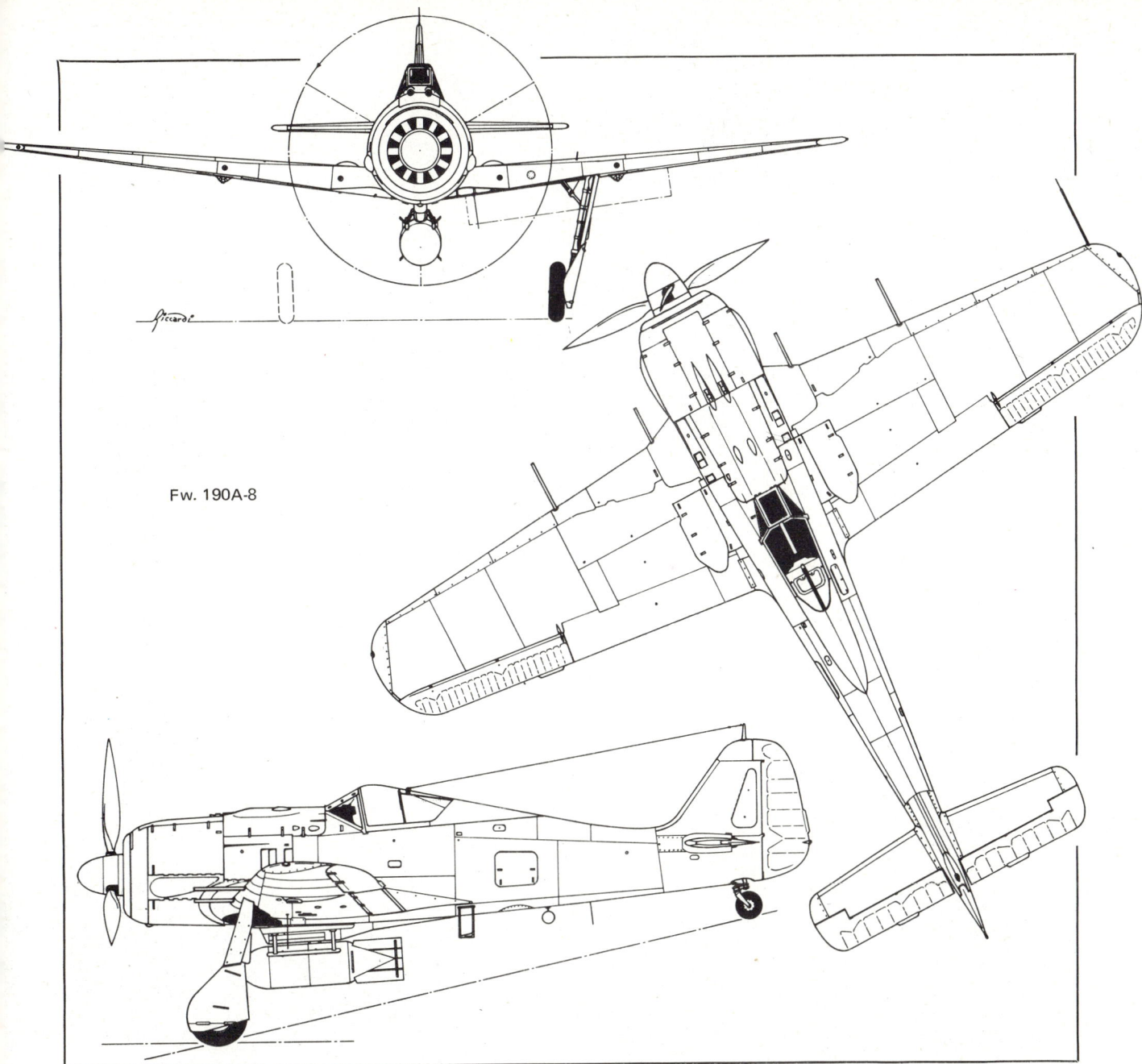

Fw. 190A-8

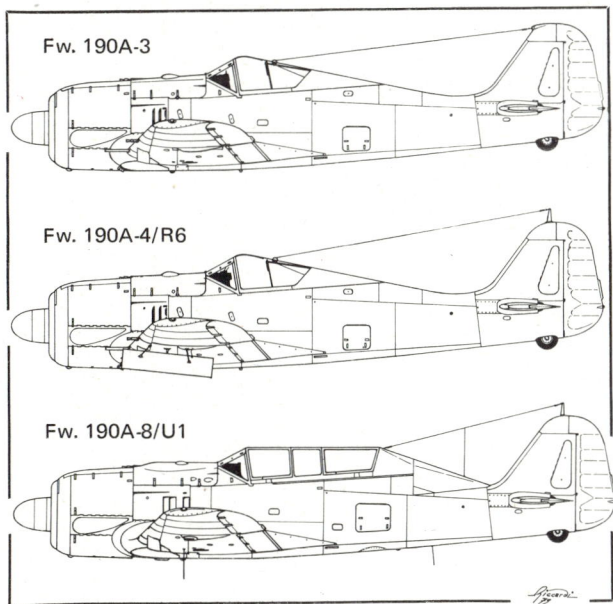

Fw. 190A-3

Fw. 190A-4/R6

Fw. 190A-8/U1

Line-up of Fw 190F-2 of 5/SchG.1 at Deblin-Irena.

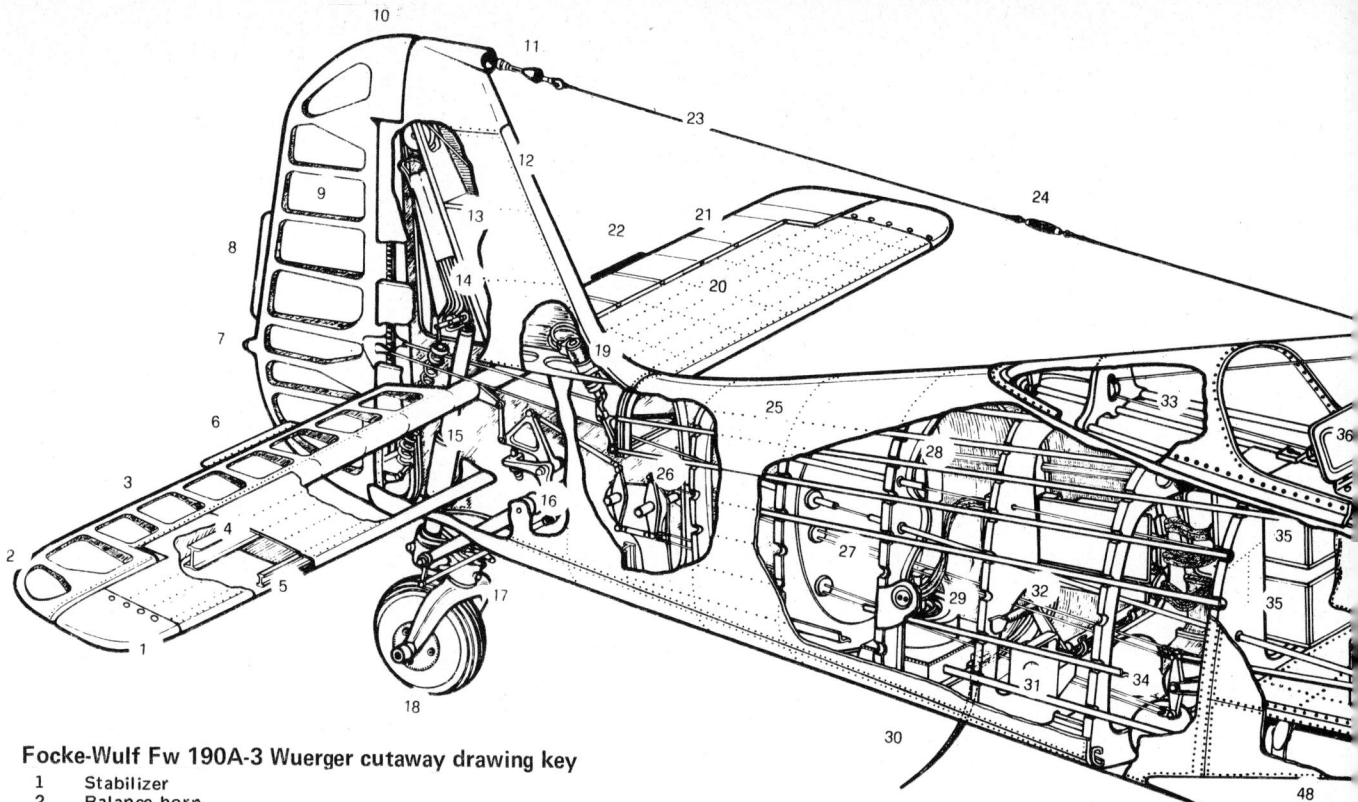

Focke-Wulf Fw 190A-3 Wuerger cutaway drawing key

1 Stabilizer
2 Balance horn
3 Elevator construction
4 Main spar
5 Front spar
6 Fixed trim tab
7 Navigation light
8 Fixed trim tab
9 Rudder construction
10 Balance horn
11 Antenna wire attachment
12 Fin
13 Fin construction
14 Tailwheel shock absorber housing
    (when tailwheel is retracted)
15 Tailwheel leg and shock
    absorber
16 Tailwheel lock
17 Tailwheel fork
18 Tailwheel
19 Tailplane incidence
    actuator
20 Tailplane
21 Elevator
22 Fixed trim tab
23 Antenna
24 Antenna wire turnbuckle
25 Fuselage skin
26 Control square
27 Fabric frame
28 Fuselage construction
29 Master compass
30 Whip aerial

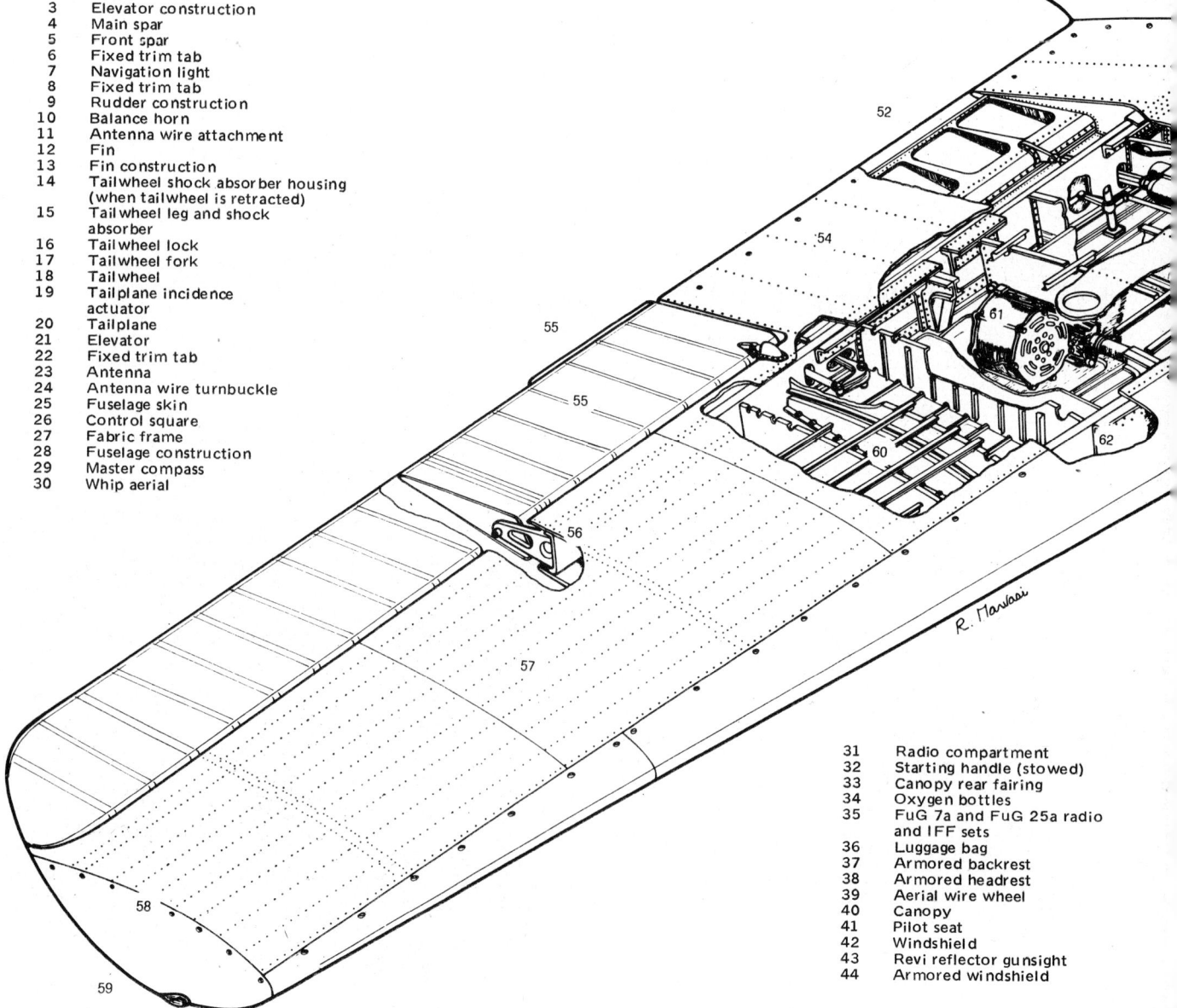

31   Radio compartment
32   Starting handle (stowed)
33   Canopy rear fairing
34   Oxygen bottles
35   FuG 7a and FuG 25a radio
      and IFF sets
36   Luggage bag
37   Armored backrest
38   Armored headrest
39   Aerial wire wheel
40   Canopy
41   Pilot seat
42   Windshield
43   Revi reflector gunsight
44   Armored windshield

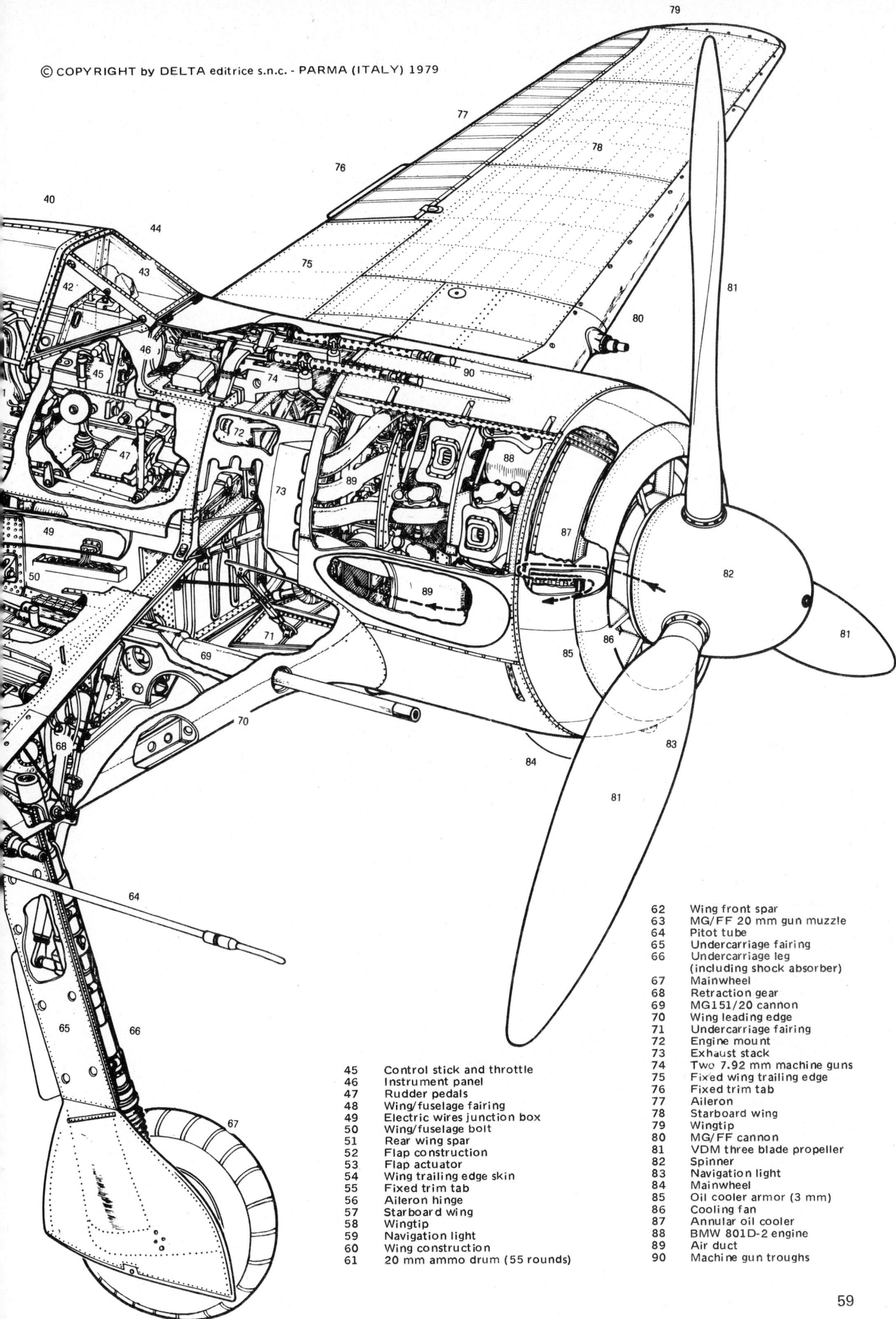

45	Control stick and throttle
46	Instrument panel
47	Rudder pedals
48	Wing/fuselage fairing
49	Electric wires junction box
50	Wing/fuselage bolt
51	Rear wing spar
52	Flap construction
53	Flap actuator
54	Wing trailing edge skin
55	Fixed trim tab
56	Aileron hinge
57	Starboard wing
58	Wingtip
59	Navigation light
60	Wing construction
61	20 mm ammo drum (55 rounds)
62	Wing front spar
63	MG/FF 20 mm gun muzzle
64	Pitot tube
65	Undercarriage fairing
66	Undercarriage leg (including shock absorber)
67	Mainwheel
68	Retraction gear
69	MG151/20 cannon
70	Wing leading edge
71	Undercarriage fairing
72	Engine mount
73	Exhaust stack
74	Two 7.92 mm machine guns
75	Fixed wing trailing edge
76	Fixed trim tab
77	Aileron
78	Starboard wing
79	Wingtip
80	MG/FF cannon
81	VDM three blade propeller
82	Spinner
83	Navigation light
84	Mainwheel
85	Oil cooler armor (3 mm)
86	Cooling fan
87	Annular oil cooler
88	BMW 801D-2 engine
89	Air duct
90	Machine gun troughs

An Fw 190A-4/U8 which landed in error at Pembrey (Engl) on 23rd June 1942. The plane carries a ventral rack ETC 501 with a SC250 250 kg bomb and two 300 liter Focke-Wulf tanks.

canopy with increased armor to protect the pilot. Generally the Fw 190s built as specialised fighter-bombers omitted the two outboard wing cannon and instead were made heavier with armor on the lower, more vulnerable areas. Even in these cases, the sub-variants possible were numerous, and similar machines were even produced by conversion of Fw 190A airframes in the field. At the end of the War, several hundred 'short nose' Fw 190s were still operational and had taken part in operation 'Bodenplatte' on 1 January 1945, which saw the launching of 1000 German fighters of various types on Allied positions. This attack caused vast destruction but also meant suffering heavy losses. Other Fw 190s were used along with the Junkers Ju88 'Kamikaze' pilotless planes, in the combination known as the 'Mistel', in which the pilots of the fighters each released his entire bomber, full of explosives, on the targets. The 'short-nosed' version of the Fw 190 with BMW 801 radial engines had shown important weaknesses in medium and high altitude performance, and still in the early stages of its development, adaptations of the airframe to take engines more suitable for high altitude use had been investigated. Only a few prototypes of these new versions, the B and C, were built, using the DB 603 in-line engines with turbo-superchargers and pressurised cabins. But these proposals were not followed up, the Fw 190D version (an adaptation of the Jumo 213A to a Fw 190A airframe), being preferred.

The first '190 to test the new installation was the V17 prototype which flew in March 1942. Apart from the new nose section, which earned it the nickname of 'Long-nose', the new fighter differed from the previous models in the lengthening of the rear fuselage for c.g. reasons. The armament consisted of two 13 mm MG131 guns in the nose and two 20 mm MG151 cannons in the wing roots.

Other prototypes followed, then as usual the D-O pre-production series and finally mass production was begun on the D-9 variant continuing the numbering of the A version (which had reached A-8).

The first Fw 190D-9 left the factory in August of 1944. The engine was the 1750 HP Jumo 213A-1 and the D-9 inaugurating the concept of the power egg, an independent system attached to the firewall bulkhead by four bolts. The fuselage had been strengthened, the armament consisted of the usual 2x13 mm and 2x20 mm guns and an auxiliary drop-tank or a 500 kg (1100 lb) bomb could be attached to detachable belly rack. The maximum speed at high altitude with MW 50 boost injection was 704 km/h (437 mph).

The D-9 was the most produced 'long-nosed' version of the Fw 190. It was used intensively in the last months of desperate fighting in the skies over Germany, proving itself to be a dangerous

Fw 190D-9 (Werk Nr. 210079) unidentified unit, crashed
on 2 January, 1945 near Brussels (Belgium).

Fw 190D-9, 15. Staffel, JG3. January-February 1945.

Fw 190D-9, 'Red Banner' Soviet Fleet of the Baltic Sea,
Finland, Summer 1945.

Ta 152H-1 (Werk Nr 150168), captured by Allied troops.

Ta 152H-1, Geschwader Stab, JG 301. Spring 1945.

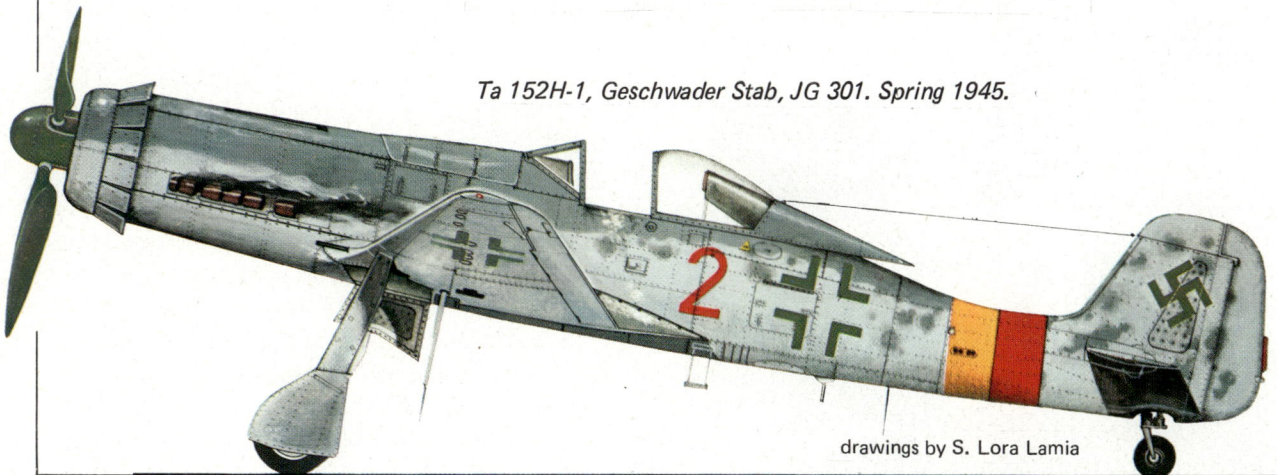

drawings by S. Lora Lamia

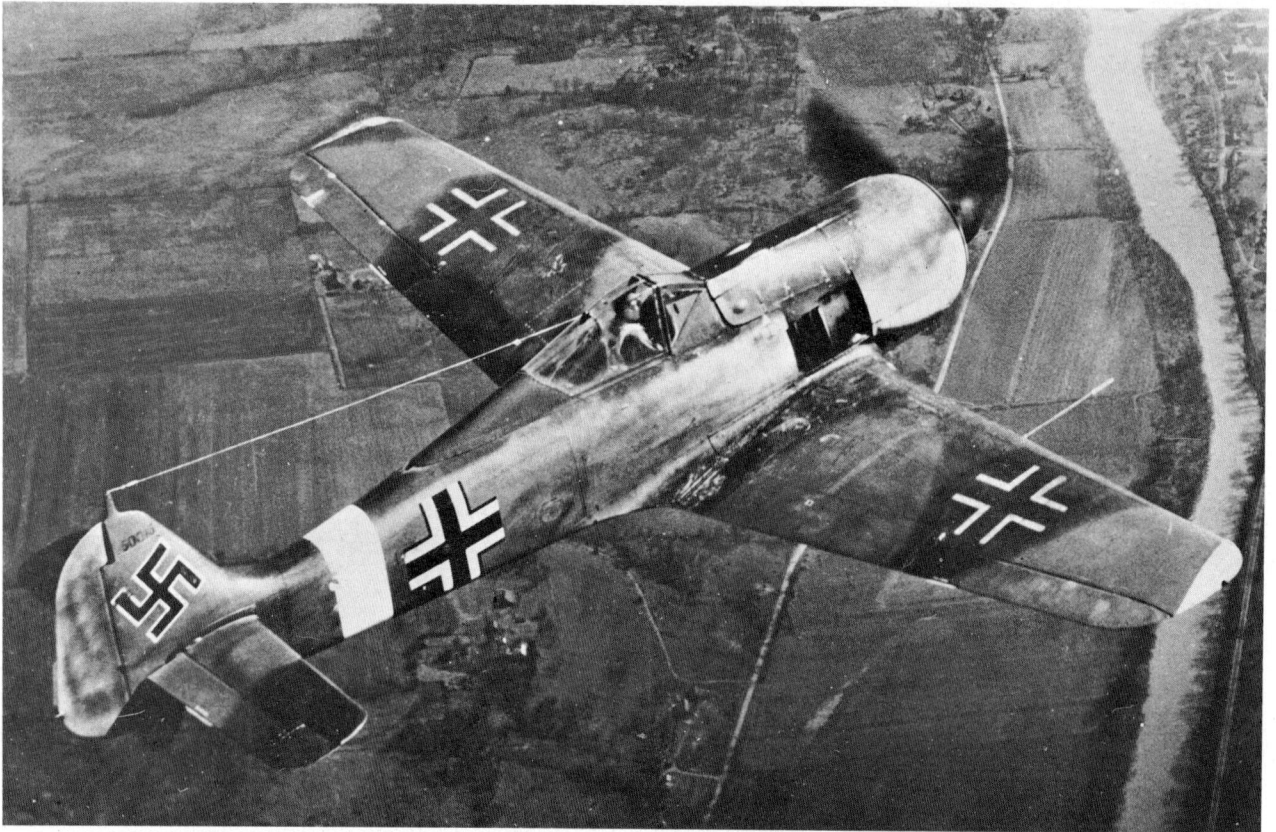

A Fw 190G-3 (Werk Nr 50045), photographed in flight in the USA after being captured.

adversary for the Anglo-American planes with which it fought. Alongside the D-9, only the Fw 190 D-12 was produced in any great quantity; it was a version adapted for ground attack, extensively armored and using a Jumo 213F engine of 2060 HP, as well as having further a 30 mm engine-mounted cannon. Its speed of 730 km/h

A Mistel S2 (trainer version) composed of a Ju 88G-1 and a Fw 190F-8.

(453 mph) at high altitudes made it an excellent interceptor, too.

The D ('Dora' in German radio code) was also used during 'Bodenplatte', alongside the 'short-noses', Me 109s and Me 262s. It was also used over England and one was brought down virtually intact. At the end of the War about 300 were still in existence, many immobilized because of lack of fuel.

But development of the successful Fw 190 airframe continued in the meantime, and the new types based on the Fw 190D were given the number Ta 152, Tank having been given permission to use his own initials 'Ta' instead of the Focke-Wulf firm's, 'Fw'. The German design engineer, therefore gave the initials Ta 152 and Ta 153 to two designs for further versions of his fighter. The first was identical to the 'Dora' except for a few details and the Jumo 213C engine with its arrangement for an engine-mounted 30 mm cannon. The Ta 153 was designed to be a high altitude fighter with the DB 603 engine, but a real prototype never flew. Information about the first versions planned, the Ta 152A and the Ta 152B, is somewhat confused but it's certain that the first production model was the Ta 152C. The function of a pre-production batch was given to a group of prototypes under the designation C-O. These were driven by 1750 HP DB 603L engines, longer than the Jumo 213A, and for which it was necessary to

A Mistel S2 composed of a Fw 190F-8 and A Ju 88G-1 probably of II/KG-200.

lengthen further the rear half of the fuselage. The armament was 4x20 mm and 1x30 mm cannons. The C-1s were substantially similar and also capable of taking the 1800 HP DB 603EM engine; it would seem that a C-3 model was also produced, with slight modifications to the armament.

The most widely built version of the Ta 152 was the H version, a high altitude fighter with longer-span wings (14.44 mm - 47 ft 4 1/2 in.), a Jumo 213 engine and a pressurized cockpit.

Other models or prototypes were also built of the H-O pre-production series, followed by a limited number of Ta 152 H-1 production models, practically identical, with the Jumo 213E of 1900 HP (2250 HP with MW50 injection). The series was terminated with a few Ta 152H-10 photographic reconnaissance planes.

70 Ta 152s were built in all, half of which succeeded in taking part in operations and most probably made the last war missions for the Luftwaffe a few hours after the surrender of Germany, their unit not having received the order for surrender. It was certainly a plane with

exceptional performance, capable of putting any adversary on the defensive. The honor for being the fastest of the Fw 190/Ta 152 family must go to the Ta 152H-1 with an exceptional maximum speed of 760 km/h (472 mph) at 12,500 m (41,000 ft). Kurt Tank himself, during a test flight of a Ta 152H, was intercepted by four American 'Mustangs' and succeeded in escaping simply by increasing speed.

The career of the Fw 190 and its derivatives thus came to an end, after being for years the plane which was always the reference point for its enemies. The same thing was true of the Ta 152, but all this was wasted effort, given the general state of disintegration of the German war machine.

Owing to the lack of precise data on German industrial production in the last months of the War, the exact production figures for the Fw 190 family are not known. They can be estimated at a total of between 19000 and 20000 models, mostly of the 'short nose' type (the 'long nose Doras ' totaled less than 700), two fifths of which were special ground attack versions.

Unlike the Me 109, the Focke-Wulf Fw 190 was not supplied to the European allies of the Germans, but for political reasons was exported only to Turkey where, during 1942, the 'Tuerk

63

At the top of the page: Fw 190D 'Long-nose'. Above: the fifth pre-production Ta 152H-0 at Cottbus.

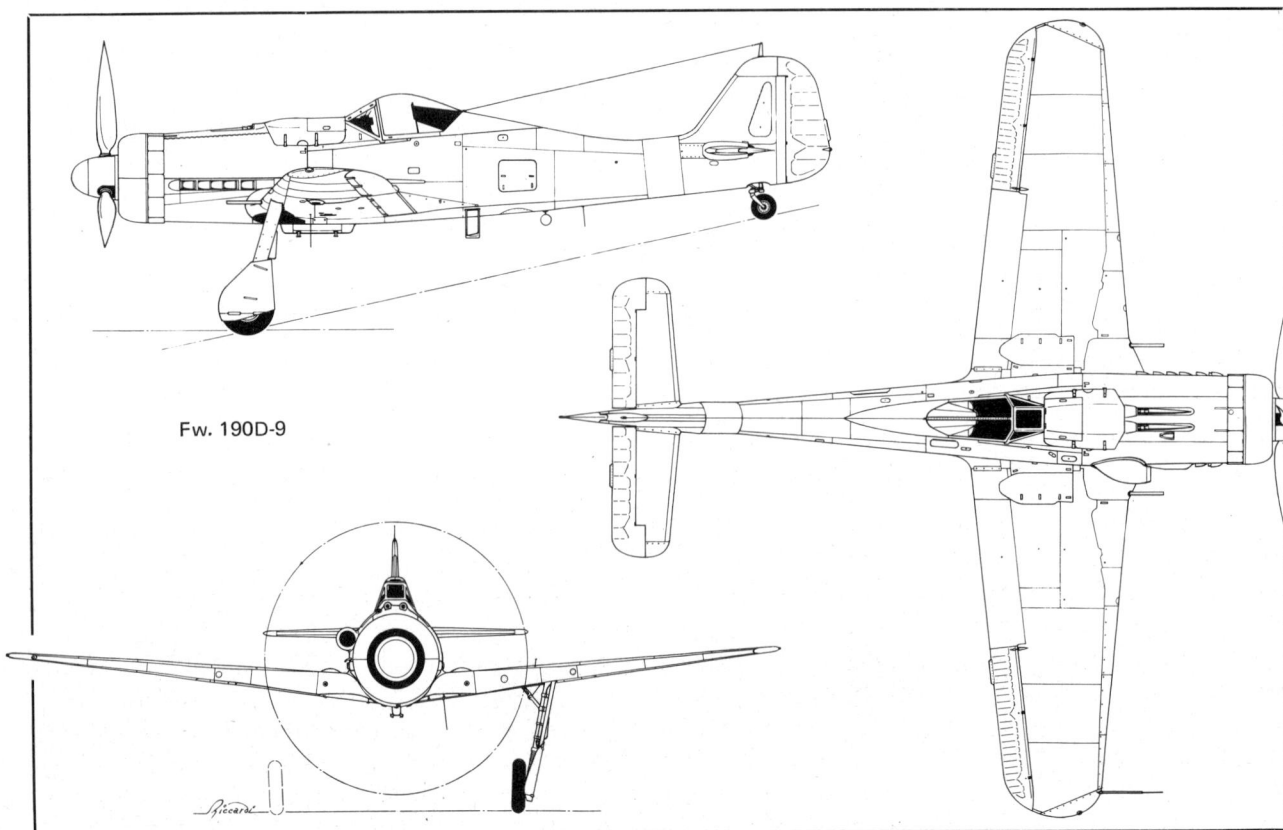

Fw. 190D-9

Hava Kuvvetlieri' received 75 Fw 190A-3 fighters. It remained in service until 1948 and was very popular with the Turkish pilots.

During the War the French works of SNCAV in Auxerre also participated in the Fw 190 production program, but only managed to deliver one before the Allied liberation. Production was resumed to re-equip the Groupe de Chasse III/5 'Normandie Niemen' of the French air force, totaling 64 Fw 190 A-8s. Unfortunately, sabotage by French workers of components built during the German occupation was such that the practical use of the plane was seriously limited, and it was withdrown from French service fairly quickly. In particular, the French-built BMW801

engines proved to be extremely unreliable.

One Fw 190A-5 arrived by sea in Japan in 1943 where it was extensively tested. Its construction design inspired the Kawasaki Ki-100, which married a radial engine to the airframe of a Ki-61, originally designed for a 12-cylinder inverted-V in-line engine.

It's worth mentioning in particular the fact that the Morskaya Aviatsiya (Soviet Navy Airforce) captured a batch of Fw 190D-9s and put them into service in the Summer of 1945 with the air component of the Baltic fleet 'Red Banner' with which it presumably operated till 1947/48.

Lastly several Fw 190s are still conserved in aeronautical museums throughout the world.

Focke-Wulf Fw 190A-B Wuerger

Multi-purpose fighter, single-seat

Power plant: one BMW 801D-2 14-cylinder, two-row, air-cooled radial engine rated at 1,700 HP for take-off and 1,440 HP at 5,700 m. Prop: three-blade metal variable pitch, constant speed VDM of 3.30 m diameter (some aircraft had wooden broad-chord props). Fuel capacity: standard: 524 liters plus one 115-liter fuselage auxiliary tank.
Dimensions: wing span: 10.506 m; length: 8.950 m; height (prop turning, tail raised): 3.390 m; wing area: 18.3 sq m; undercarriage track: 3.5 m.
Weights: empty equipped: 3,470 kg; loaded: 4,380 kg; max. T.O. weight: 4,865 kg.
Performance: max. speed: 656 km/h at 6,300 m (with GM-1), 647 km/h at 5,500 m (without GM-1), 571 km/h at sea level; cruise; 477 km/h; approach speed: 250 km/h; stalling speed: 204 km/h; initial climb rate: 17.5 m/sec; climb to 6,000 m: 9 min 6 sec, to 8,000 m: 14 min 24 sec, to 10,000 m: 19 min 18 sec (26 min 5 sec according to other sources) or 16 min 5 sec with GM-1; service ceiling: 10,300 m (11,400 m with GM-1); combat radius: 414 km; range: 1,035 km at 7,000 m; ferry range: 1,470-1,620 km.
Armament: two Rheinmetall-Borsig 13-mm MG131 machine-guns with 400-475 r.p.g. and four Mauser MG151/20E wing mounted cannons (the inner ones with 250 r.p.g. and the outer ones with 125-140 r.p.g.).

Focke-Wulf Fw 190D-9 (Dora)

Day interceptor with ground attack capability, single-seat

Power plant: one Junkers Jumo 213A-1, 12-cylinder, inverted Vee, liquid-cooled engine, rated at 1,750 HP for take-off, 1,600 HP at 5,500 m, 2,240 HP with MW-50 at sea level, 2,000 HP at 3,400 m (MW-50) and 1,880 HP (MW-50) at 4,725 m. VDM VS-111 three-blade, variable pitch, constant speed propeller. Fuel capacity: 523 liters plus one 113-liter optional auxiliary tank in fuselage.
Dimensions: wing span: 10.50 m; length: 10.19 m; height: 3.36 m; wing area: 18.30 sq m.
Weights: empty: 3,490 kg; loaded: 4,382 kg; max. take-off weight: 4,840 kg.
Performance: max. speed: 704 km/h at 11,300 m, 681 km/h at 6,500 m, 634 km/h at 3,000 m, 570 km/h at sea level; climb to 2,000 m: 2 min 6 sec, to 4,000 m: 4 min 30 sec, to 6,000 m: 7 min 6 sec, to 10,000 m: 16 min 48 sec; service ceiling: 10,000 m; absolute ceiling: 11,300 m; combat radius: 334 km; range: 836 km at 5,600 m; max. range: 1,120 km.
Armament: two Rheinmetall-Borsig 13-mm MG131 machine guns with 475 r.p.g. and two Mauser MG151/20E 20 mm guns with 250 r.p.g. plus one SC500 500 kg bomb.

Focke-Wulf Ta 152H-1/R11

Day and limited all-weather air superiority fighter, single-seat

Power palnt: one Junkers Jumo 213E-1, 12-cylinder, inverted Vee, liquid-cooled engine rated at 1730 HP for take-off (2,250 HP with MW-50) and 1300 HP at 10,000 m (1,715 HP with GM-1). Propeller: see Fw 190D-9. Fuel capacity: 993 liters (internal) and 300 liters (external).
Dimensions: wing span: 14.43 m; length: 10.71 m; height: 3.36 m; wing area: 23.30 sq m.
Weights: empty: 3,920 kg; loaded: 4,750 kg; max. TO weight: 5,217 kg.
Performance: max. speed: 759 km/h at 12,500 m (with MW-50 and GM-1), 748 km/h at 9,000 m (with MW-50), 672 km/h at 10,600 m, 534 km/h at sea level (563 km/h with MW-50); cruise: 439-500 km/h at 7,000-10,000 m; max. continuous cruise: 605 km/h at 10,000 m; initial climb rate: 17.5 m/sec (with MW-50); service ceiling: 14,800 m (with GM-1); combat radius: 486 km; range: 1,215-2,010 km.

Armament: one Rheinmetall-Borsig 30 mm MK108 cannon with 90 rounds and two Mauser 20 mm MG 151/20E cannons with 175 r.p.g.

HEINKEL He 219 'Uhu'

The night fighters of the Second World War were all adaptations of already existing models, the sole exception being the American Northrop twin-engined P-61 Black Widow, designed from the beginning for this rôle.

The Heinkel 219 Uhu (Owl), considered by many to be the best German night fighter in service was ordered as such from the start, but originally the project had been conceived as a 'zerstörer' (destroyer), a project so dear to the Germans.

The design for this twin-engined plane was prepared on Heinkel's own initiative in the Summer of 1940. Not responding to any specification or any immediate necessity, it was an opportunity for the engineers of the Heinkel's Technical Department to complete a brilliant exercise in designing the most technically advanced plane possible. In fact this project featured remote controlled defensive gun barbettes, a pressurized cockpit, retractable tricycle landing-gear, and, on the final production model, for the first time in history, ejection seats for both crew members. Understandably Heinkel's proposal did not meet with government approval, because of the excessive number of innovations introduced and, also, the guaranteed availability of existing planes with the necessary characteristics.

The He 219 was resurrected when General Kammhuber, who was looking for a night fighter to oppose the potentially very dangerous four-engined offensive bombers of the RAF, began to be interested in developing this twin-engine design as a specialized night fighter.

Above: He 219A-O/R-1 captured and taken to America. Below: He 219 V-1 prototype converted to He 219A-O captured by RAF.

For its entire life span the He 219 found itself at the center of a vast complex of jealousies and rivalry between its supporters and critics in the heart of the Air Ministry and other organizations responsible for its development.

Most of 1942 was spent in detail design and the construction of the prototype under the auspices of Kammhuber, who was ansious to have the promising nocturnal 'bird of prey' as soon as possible. During these months the RLM rejected the idea of remote-controlled barbettes, and work was delayed by a bombing raid on Rostock which forced Heinkel to transfer the program to Vienna Schwechat.

The first prototype, the He 219 V1 flew on 15 November 1942, powered by two 1750 HP Daimler-Benz DB 603A engines in place of the DB 603Gs originally planned, as the latter were still not fully developed. Its flying characteristics were completely satisfactory. Other prototypes immediately followed, and various types of armament combinations were tried out, using 15, 20, and 30 mm cannons, all mounted in a ventral tray. The elimination of the remote controlled barbettes led to the gradual modification of the fuselage contours and to the adoption of a 13 mm machine-gun on a flexible mount. From the He 219V4 on, an FuG 212 'Lichtenstein' C-1 radar system was installed.

The RLM, which had already ordered 100 models before the first flight, increased the quantity to 300. Production was set up at Schechat, but also using components from Milec and Buczin in Poland and also from the sections of Rostock-Marienehe remaining active.

The pre-production A-O models, as was the custom, were given the name 'Versuchs' (experimental) and some of these were used for service testing, based at Venlo in Holland alongside the other night fighters of I/NJG 1.

The first production model should have been the A-1 with the DB 603E engine, but the unavailability of this powerplant made it necessary to retain the DB 603A and so the designation became A-2. The armament was composed of 2x20 mm guns in the wing roots and 2x15 mm (or 2x20 or 2x30 mm, depending on availability) in the ventral tray. The useless dorsal defensive weapon disappeared completely. The two 30 mm guns for oblique upwards firing 'Schräge-Musik' (Jazz music) were not always installed at the factory, but often they were fitted once the aircraft reached their chosen units. After the first models were complated with FuG 212, the Germans used the new FuG 220 'Lichtenstein' SN2, operating on a frequency of 490 megacycles was fitted, the British having adopted effective countermeasures against the 90 megacycles band of the preceding model from the Summer of '43.

The production rate of the He 219 was very low for a long time, so low, in fact, as to give its enemies a good excuse for demanding its suspension. This did not happen, however, as the other types more favored by the RLM were also behind schedule (the Ju 88G, for example). For this reason, it was necessary to bring into service whatever was available and even more so with the He 219 because of its excellent performance.

To make this plane more acceptable to the RLM, who were still conditioned by old-fashioned notions on multipurpose twin-engined aircraft, Heinkel submitted A-3 and A-4 bomber versions, but without success.

The few He 219s in service had, in the meantime, given excellent results, the most famous of which was the first mission from Venlo, in which Major Streib, the commander of 1/NJG 1 brought down 5 bombers. As if this was not enough, in the next six missions, the 'Uhus' of Venlo shot down twenty RAF bombers, including six Mosquitos.

In March 1944 the first He 219A-5s appeared; they differed little from the A-2. Different sub-series of the A-5 were given, according to availability, 1900 HP DB 603Gs instead of the usual DB 603As. Some models, designated He-219 A-5/R4, were given a defensive weapon fired by a third crew member in an attempt to hinder the Mosquito night fighters which accompanied the English bombers in ever increasing numbers, to protect them from the German night fighters.

Numerous sub-versions followed on the A-5 and with the reliable availability of the DB 603G engine, the final and most widely used version was put into production, with the designation He 219A-7. This type introduced improvements in equipment, in the protection of the crew and in having more powerful weapons. An FuG 218 'Neptun' radar was coupled with the FuG 220 and in many cases there were 4 guns in the ventral tray in addition to the two in the wing-roots which were changed from 20 mm to 30 mm.

In November 1944 a Fighter Emergency Program was issued by the RLM which sanctioned the suspension of production of various aircraft, among them the He 219. The order was, however, tacitly ignored and although reduced, production went on to supply the units which already had this plane in service. The He 219 was so greatly appreciated that it led to a curious episode in which six models were constructed entirely by a unit itself using spare parts, and consequently were put into service with no Werk-Nrs.

The He 219 B with its greater wing span should have followed the A version, but only a few prototypes were built for testing the different versions projected. The He 219B-1 was the same as the A-5/R4, but was to have had Jumo 222/B-3 engines, though the prototype flew with the usual DB 603As. The He 219 B-2 was a lighter 'anti-Mosquito' version, derived from the He 219A-6. A few models were built using DB 603L engines, but the armament was too light without the ventral fuselage cannons. This was followed, therefore, by the He 219B-3, more heavily armed and supposed to have the Jumo 222 engine which, however, never arrived. The evolution of the 'Uhu' concluded with the He 219C, whose fuselage was extensively modified and for which both a fighter and a bomber version were planned, both to be armed with tail turrets for four 13 mm machine guns. Neither was never built.

Heinkel He. 219A-7/R1 Uhu

High altitude night interceptor fighter, two-seat

Power plant: two Daimler-Benz DB-603G, 12-cylinder, inverted Vee, liquid-cooled engines, rated at 1,900 HP for take-off. Three-blade, variable pitch, metal propellers.
Dimensions: wing span: 18.50 m; length: 15.52 m; height: 4.10 m; wing area: 44.5 sq m.
Weights: empty: 11,200 kg; loaded: 15,300 kg.
Performance: max. speed: 670 km/h at 7,000 m; max. cruise: 630 km/h; normal cruise: 540 km/h; initial climb rate: 9.1 m/sec; service ceiling: 12,700 m; range: 1,545-2,000 km.
Armament: two Mauser 20-mm MG151/20A cannons and two Rheinmetall-Borsig 30 mm MK103 cannons in a ventral tray, two Rheinmetall-Borsig 30 mm MK 108 cannons in the wing roots and two other MK108s in an oblique-firing 'Schraege Musik' installation.

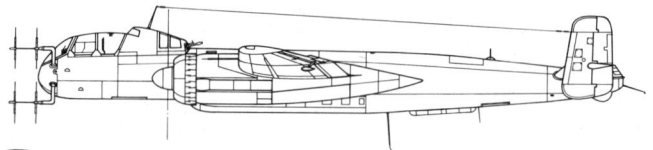

He. 219A-5/R-4

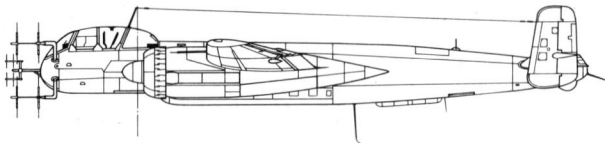

He. 219A-5/R-1

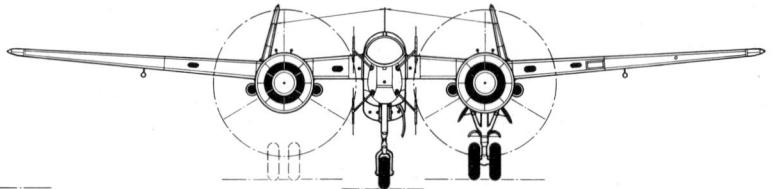

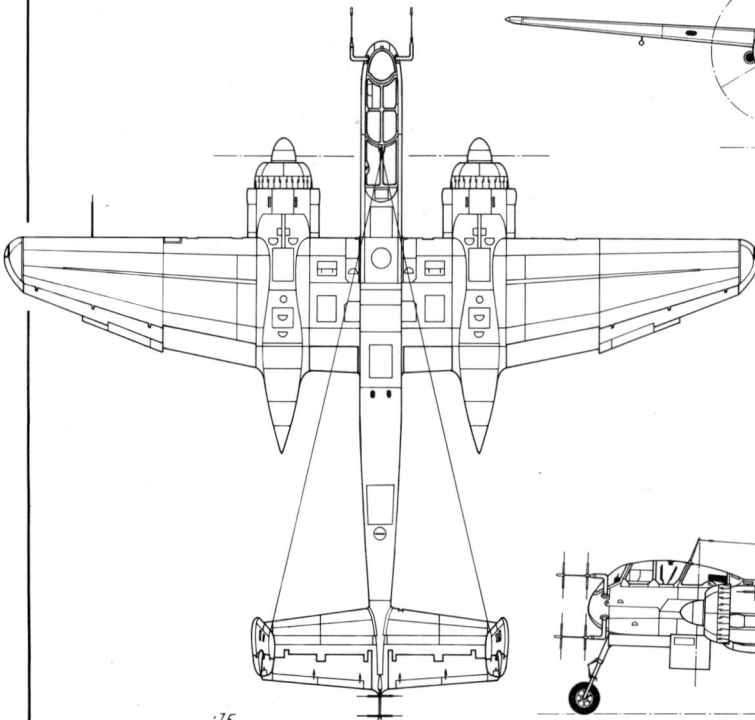

He. 219A-7/R-4

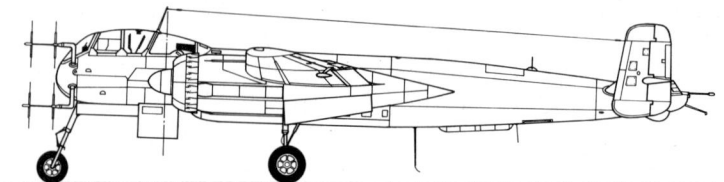

HEINKEL He 162 'Volksjäger'

He 162A-2 'Volksjäger' in flight.

Notwithstanding the precarious conditions in which Germany found itself at the end of 1944, at the height of the Allied air offensive, its industry managed to maintain a surprising vitality under the guidance of the Armaments Minister of the Reich, Albert Speer.

The most extraordinary demonstration of this vitality is the history of the Heinkel He 162 Volksjäger (People's Fighter) created from a specification issued by the RLM on 8 September 1944 and flown for the first time on 6 December of the same year! At the end of the War, more than 100 had been built, many of which were in fact delivered to operational units, even if they do not seem ever to have been used in combat.

The specification was even more demanding, requiring an initial production rate of 1000 aircraft a month, to be ready in the first half of 1945! The general concept that inspired the specification was that of a 'light fighter': a simple plane to manufacture, furthermore using non-strategic materials, and having a high performance level, but suitable to be flown by semi-trained personnel and pilots. The last two requirements were clearly contradictory, but the principal problem facing the Germans in the last phase of the War was precisely that of training, and they even had the absurd idea of trusting the new planes to boys of the Hitlerjügend, after brief training on gliders!

The He 162 project, finally approved on 30 September, was developed in detail at the same time as the first prototype He 162 V1 was being built. The plane was called the 'Salamander', but was nearly always known as the 'Volksjäger'. The fuselage was a metal monocoque while the wings were wood, with plywood covering. Many sub-contractors participated in the production program, and parts of the assembly lines were set up in caves and mines to protect them from air attacks. The BMW 003A of 750 kg st (1760 lb st) was mounted on top of the fuselage and the narrow-track under carriage was situated entirely in the fuselage.

After the first flight, concluded successfully, the first prototype was demonstrated in flight to the political and military authorities interested in the 'Volksjäger' program, but during a low level pass the defective bonding of the right wing components resulted in the wing disintegrating, and the plane was destroyed. However, the program went on and mass production was in full swing during the final development of the

This photo shows anhedral wing-tips of the He 162A-2. The tail was of mixed duralumin, steel and wood.

prototypes. The only variation worth mentioning, which was introduced on the subsequent prototypes and kept on the production models, was the use of anhedral wing-tips.

Feverishly, the He 162A-O models were put into production. They all received the 'Versuchs'

He 162A-2

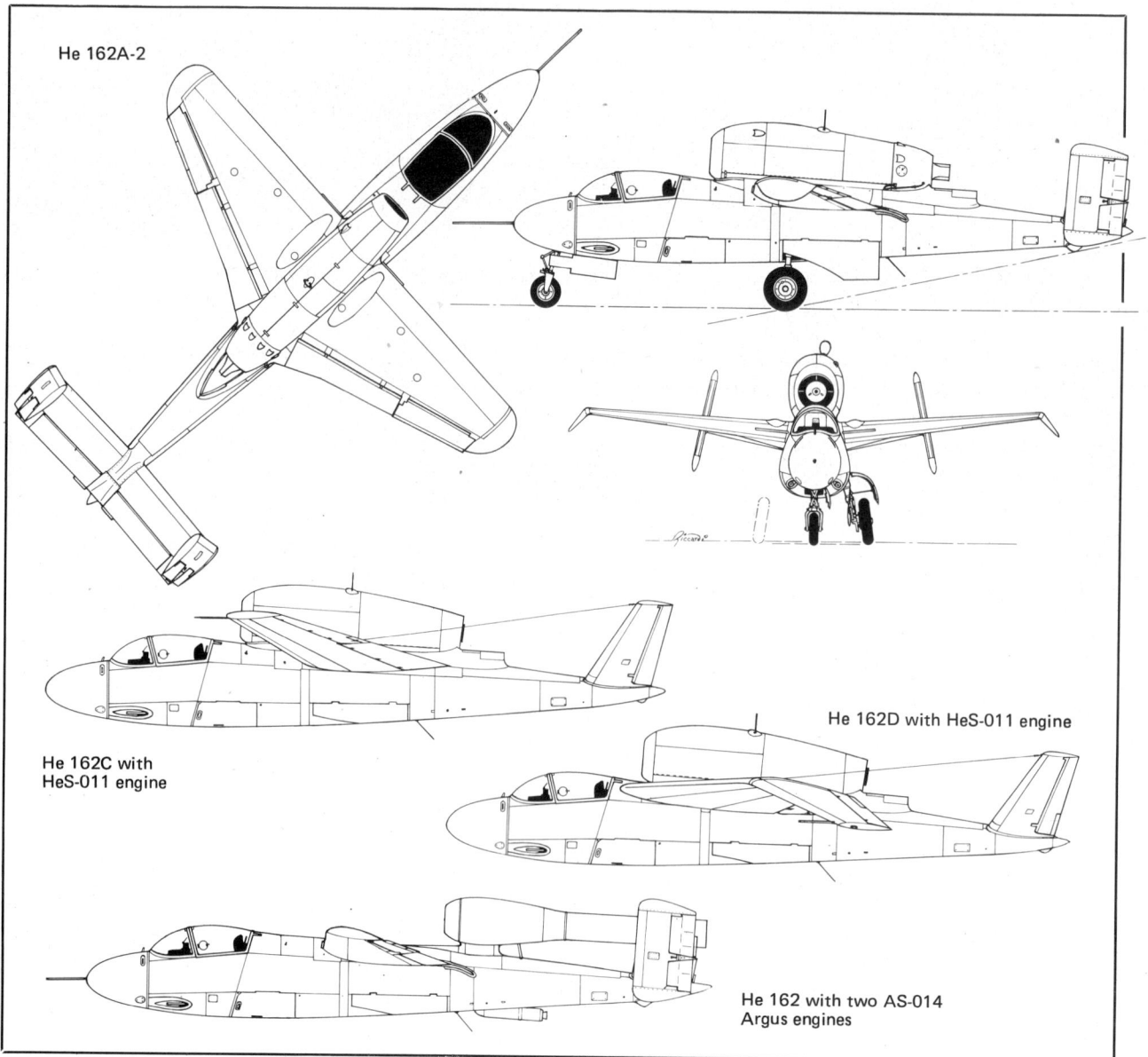

He 162C with HeS-011 engine

He 162D with HeS-011 engine

He 162 with two AS-014 Argus engines

(experimental) designation and were built with the A-1 and A-2 versions, which were substantially similar and armed with 2x20 mm guns. The A-3 was fitted with the heavier Mk108 30 mm cannon.

As with all the jet planes of the Luftwaffe, a specific evaluation unit was created, called EK 162, but the situation was such that its duties overlapped those of the first operational unit, I Gruppe of JG 1. In the chaotic situation of those months, this unit and the others which were hurriedly formed, tramped from one airfield to another without being able to perform their proper duties and with occasional but bloodless encounters with Allied aircraft.

Basically the He 162 proved to be a well-made plane even if inadequately developed. Certainly it was fortunate that it arrived too late to be put into the hands of the adolescents of the Hitler Youth groups: being little and light did not necessarily make it an easy plane to pilot and take

into combat, and Galland was probably right when he protested about this waste of energy which would have been better used on production of the already tested and developed Me 262.

In spite of its brief life, the He 162 lent itself to a considerable number of changes in form, and other experimentation (the majority of which never materialised or were still being set up at the end of the War). The use of alternative power plants had been planned, such as the Heinkel-Hirth 011 and the Argus pulse-jet engine. For future versions the adoption of swept-back and swept-forward wings, were to be decided by testing, and a 'butterfly' tail assembly was planned. The fastest version planned was the E model, with a BMW 003R engine, the mixed power plant formed by the coupling of a normal BMW 003 turbojet engine with a liquid-fuel BMW 718 rocket motor, capable of supplying a supplementary thrust for climbing and in emergencies.

Heinkel He 162 A-2

Single-seat interceptor fighter

Power plant: one BMW 109-003 E-1 rated at 800 kg thrust.

Dimensions: span 7.20 m; length 9.05 m; height 2.60 m; wing area 11.16 m^2.

Weights: empty 1663 kg; empty equipped 1758 kg; max. loaded 2907 kg.

Performance: maximum speed (with short period max. thrust of 920 kg) 905 km/h at 6000 m, 890 km/h at sea level; climb rate (normal power) 19.2 m/sec at sea level, 9.9 m/sec at 6000 m; ceiling 12000 m; range at full throttle 620 km at 6000 m.
Armament: two 20 mm Mauser MG151 cannon with 120 rpg.

DORNIER Do 217 night fighters

The major problem of the Luftwaffe, during the first stage of organizing night fighter units to defend the Reich from RAF bombers, was that of making available sufficient planes suitable for the new rôle. They were forced to resort to adaptations of types already in existence. The most highly thought of was the Junkers Ju 88 for its brilliant performance, heavy armament and long endurance, which permitted long standing patrols. The scarcity of Ju88s available, however, forced the Luftwaffe to turn to the Bf110 which at first was the most widely used version and carried out its tasks well. But the need for night fighters was still not satisfied, and so a specialized version of the Do 217E twin-engined bomber, designated Do 217J, had to be adopted. The plane revealed various weaknesses as a night

fighter and was certainly the least appreciated type of aircraft among night fighter pilots. Its adoption was due to the fact that similar conversions had been done from 1940 on its predecessors, the Do 17Z and Do 215, for service as intruders and night-fighters, using the infra-red 'Spanner Anlage' sensors. These first conversions of the Dornier bombers to heavy night fighters were produced in only small quantities, as the Ju88 was preferred even then. However, the few planes in service played a successful rôle, especially in trailing and attacking English bombers during their landings on returning from offensive missions. The Dorniers were less successful in defensive roles as they were not fast enough. The Do 17Z-10 and the Do 215B-5 differed from one another only in the engine used: the 1300 HP

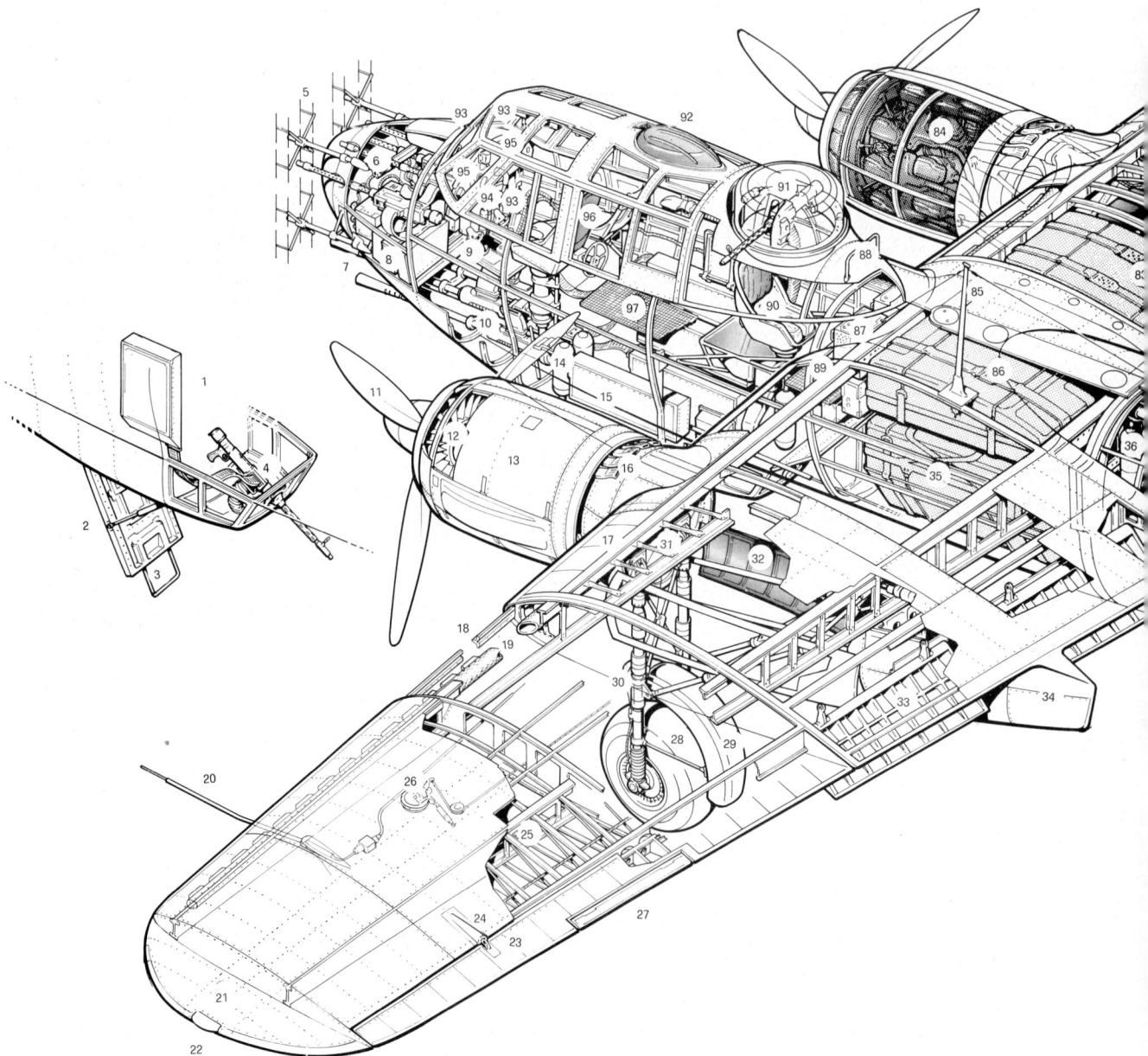

72

Dornier Do 217J-2 cutaway drawing key

1. 'C' station: machine gun often deleted for night missions
2. Entry hatch
3. Foot step (retracting)
4. MG131 machine gun
5. FuG 202 Lichtenstein radar with 'Matratzen' aerials
6. Four MG 131 machine guns
7. MG/FF gun flame dampers
8. 13 mm ammo boxes (4,000 rounds)
9. Rudder pedals
10. Four cannons (two MG/FF and two MG-151/20A)
11. VDM three blade propeller
12. Cooling fan
13. Engine cowling
14. Compressed air bottles (for MG151 re-cocking)
15. 20 mm ammo boxes (1,400 rounds)
16. Exhaust stacks
17. Wing leading edge
18. Wing leading edge false spar
19. Hot air de-icer
20. Pitot tube
21. Wingtip
22. Navigation light
23. Aileron
24. Aileron hinge
25. Wing rear spar
26. Retracting landing light
27. Trim tab
28. Mainwheel
29. Mudguard
30. Undercarriage leg
31. Wing front spar
32. Undercarriage door
33. Flap (lowered)
34. Engine nacelle tailcone
35. 1,160 liter auxiliary tank
36. Compressed air bottles
37. Oxygen bottles
38. Dinghy
39. Treadway (catwalk)
40. Fuselage frame
41. VHF aerial
42. Master compass
43. Batteries
44. Fuel vent
45. Tail wheel retracting rod
46. Mudguard
47. Tailwheel
48. Fin leading edge slots
49. Fin
50. Balance horn
51. Rudder
52. Trim tab
53. Tailplane
54. Fixed trim tab
55. Fuel vent
56. Ballast
57. Tail cone
58. Navigation light (white)
59. Stabilizer (right)
60. Fixed trim tab
61. Spar
62. Stabilizer hinge
63. Stabilizer construction
64. Fin construction
65. Rudder
66. Rudder trim tab
67. Fuselage Construction
68. Aerial
69. Compressed air bottles
70. Flap construction (lowered)
71. IFF FuG.25a Transponder
72. FuG. 101 radioaltimeter
73. Aileron construction
74. Aileron trim tab
75. Navigation light (green)
76. Trailing edge spar
77. Rear spar
78. Front spar
79. Leading edge de-icing system
80. 160 liter tank
81. 235 liter tank
82. Flap actuator
83. 795 liter tank
84. BMW 801ML engine
85. Aerial mast
86. 1,050 liter tank
87. Radio sets
88. Gun deflector arc
89. Floor
90. Observer/gunner seat
91. Upper turret (often without machine gun for night missions)
92. D/F dome
93. Armor glass
94. Control wheel and stick
95. Instrument panel
96. Armored pilot's seat
97. Floor

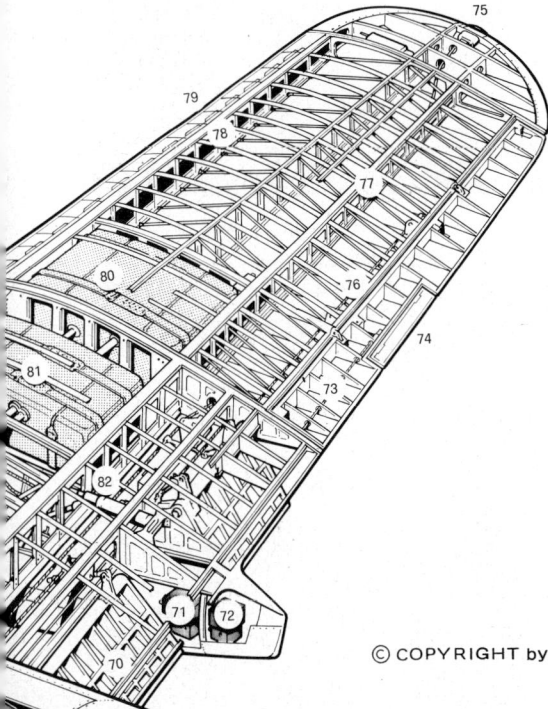

© COPYRIGHT by DELTA editrice s.n.c. - PARMA (ITALY) 1979

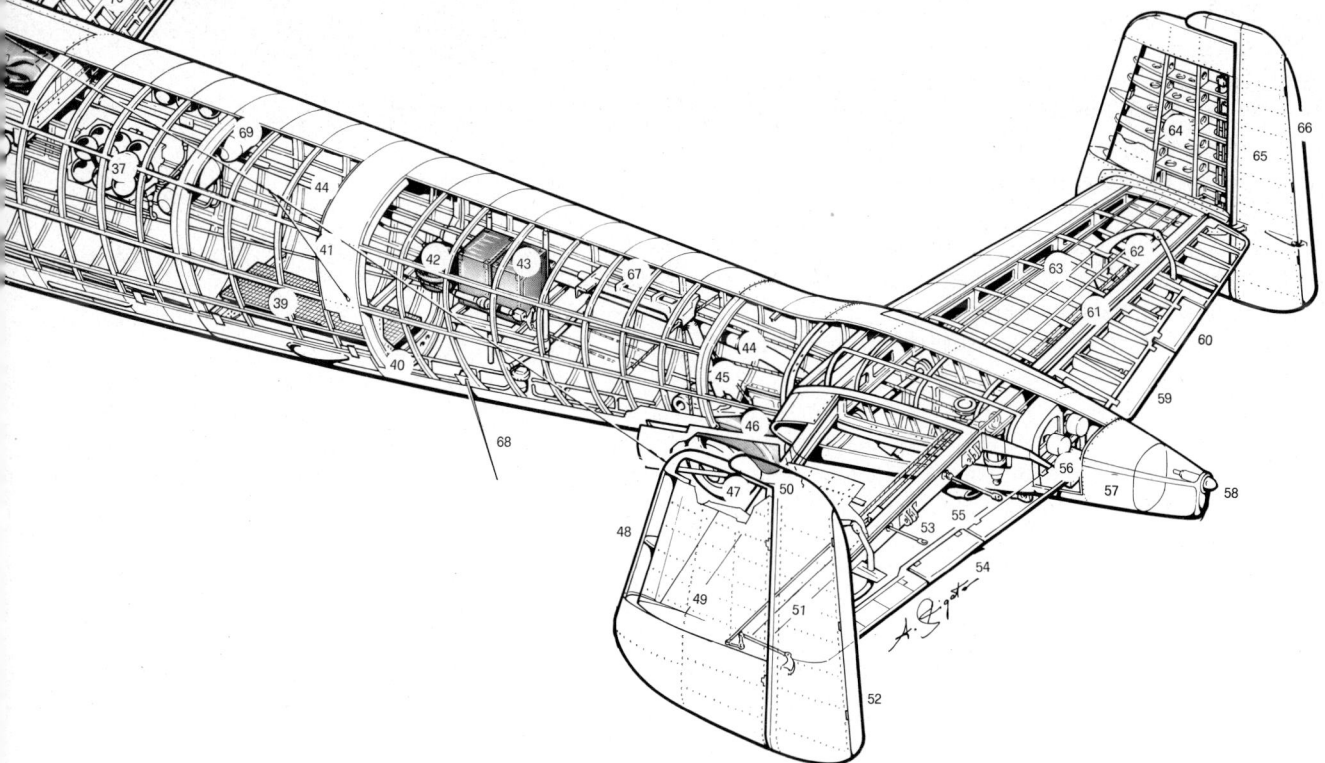

73

BMW 323R radial for the former and 1075 HP Daimler-Benz DB 601A in-line engine for the latter. Both these planes were available early enough to be able to use (in some cases) the first airborne radar system, and it seems that they were the first German planes to use the fixed weapons for oblique upward firing 'Schräge Musik' (Jazz music).

In 1942, they were replaced by their heavier and more powerful successor. This plane, the Do 217, was more widely used, too, but was even less popular with the crews. The first Do 217s were converted from existing airframes and were designated J-1. They had no radar and carried an extremely powerful armament of 4x7.9 mm and 4x20 mm guns in a redesigned nose. In addition, they kept their bombing equipment and the capability to carry out bombing missions.

More radical modifications were introduced with the Do 217J-2: the elimination of the bomb bay and the adoption of an FuG 202 'Lichtenstein' BC and later the FuG 212 'Lichtenstein' C-1 radar system (the latter was the same as the other but more suitable for mass production).

Development of the plane continued and the Do 217N version was adapted from the Do 217M bomber. The engines were the main difference between the J and N versions, being those of the respective bomber versions from which they were derived: the first had 1600 HP BMW 801ML 14 cylinder radial engines, while the Do 217N used the Daimler-Benz DB 603A, 12-cylinder in-line engines of 1750 HP.

The Do 217N-1 was practically the same, except for the engine, as the J-2, but retained the aft bomb bay as in the J-1, being designed secondarily as an intruder.

The Do 217N-2, on the other hand, omitted all

From top to bottom: a Dornier Do 217J-1 of the Regia Aeronautica. A Do 217J-2 of the Luftwaffe. A Do 217 N-2 without radar aerials.

bombing equipment as well as the defensive armament and was the most 'specialized' night fighter in the family. It was also the final model of the series, being phased out of production at the end of 1943, when the output of night fighters more suited to the rôle was sufficient for the requirements of the Luftwaffe.

When the N-2 came into service, the radar system was changed from the FuG 212 'Lichtenstein' C-1 operating on 90 megacycles to the FuG 220 'Lichtenstein' SN2 transmitting on 490 megacycles. This was because the RAF had found an efficient countermeasure to the former frequency band in the form of the so-called 'Window', tiny metal strips of suitable dimensions, released by the British bombers and having the effect of multiplying the image of the

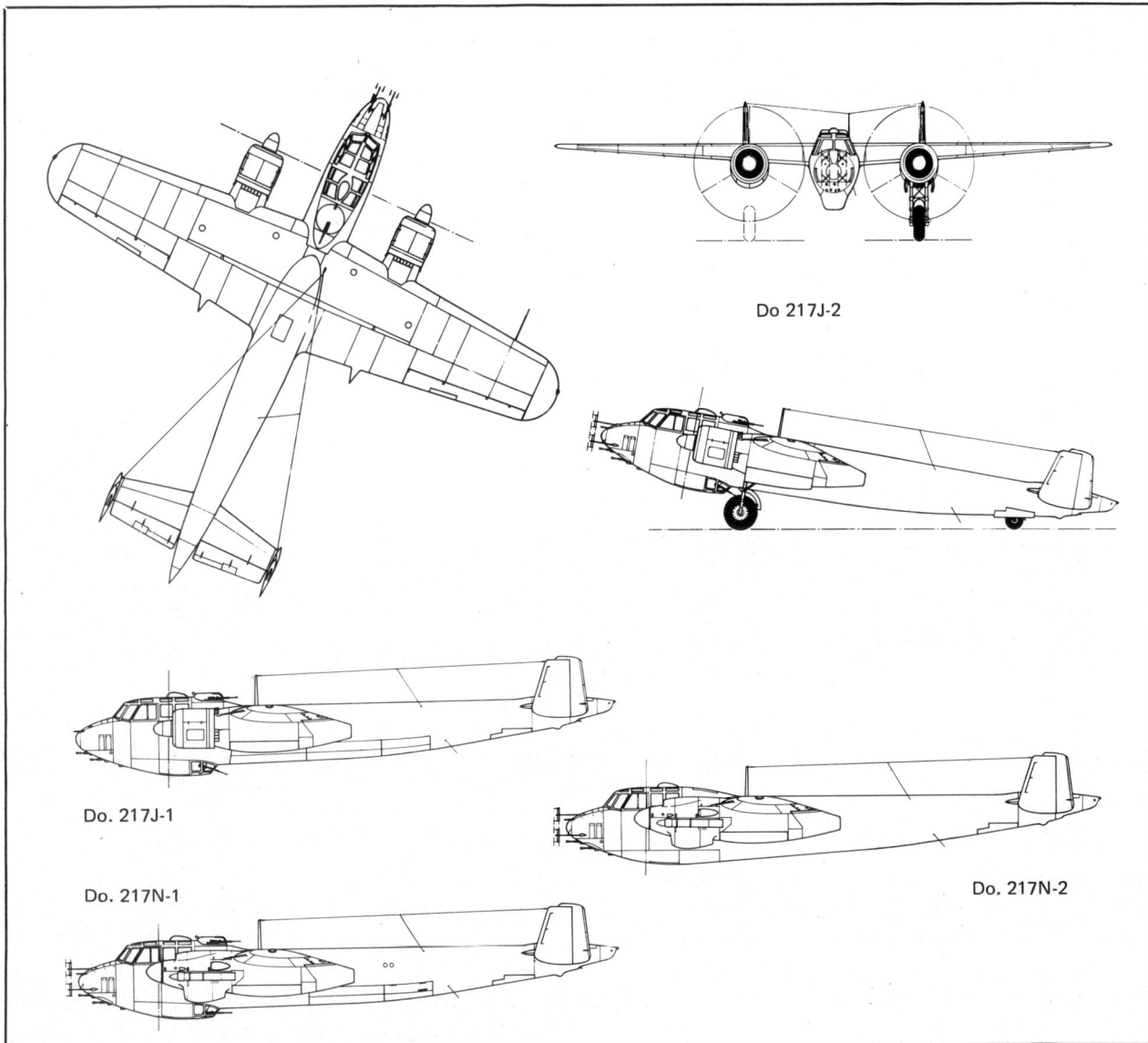

Do 217J-2

Do. 217J-1

Do. 217N-1

Do. 217N-2

attacking aircraft on the surface-based 'Wurzburg' radar screens and those of the 'Lichtenstein' C-1 radar sets in the fighters.

Frequently both types of radar were kept on the Do 217 (as on all the other German night fighters) in that the new FuG 220 'Lichtenstein' SN2 was effective only down to a minimum distance of 400 m while the FuG 212 'Lichtenstein' BC was capable of keeping contact with the target down to 200 m.

Only a few hundred Do 217 night fighters were built; as has already been said, it was the least appreciated of the types performing that particular rôle at the time, but nevertheless it was a useful supplement for the men of the Nachtjagdgeschwadern of the Luftwaffe.

In Italy the Regia Aeronautica also had to face the problem of night-fighter units, but did not have at its disposal either the means or, more important, the organization needed. The Italians turned to their ally, Germany, who gave them the idea of forming a 'Comando Intercettori'

organization with adequate equipment and agreed to help train personnel, this taking place at Venlo in Holland in the Summer of '42, using Do 217s. They also agreed to supply Bf 110s but the Italians did not receive the quantities required (only 3 received the insignia of the Regia Aeronautica) and so the Luftwaffe gave them 12 Do 217J-1s and 8 Do 217J-2 fighters, which were used (with scant success) till the Italian armistice. Subsequently the surviving planes were taken back into service by the Luftwaffe. Not all the Italian Do 217J-2s had radar. To substitute for the Do 217, the Regia Aeronautica had ordered a special version of the Cant. Z 1018 bomber with 'Lichtenstein' SN2. The Italian plane was of the same class in weight and performance as the German twin-engined design.

Both the N-1 and the N-2 used batteries of two or four 20 mm cannons firing obliquely upward from the top of the fuselage. This installation was called 'Schräge Musik' ('Jazz music') by the Germans.

JUNKERS Ju 88 night fighters

Ju88R-1 in flight.

One of the most outstanding achievements of the aeronautical industry in Germany immediately before the war was supplying the Luftwaffe with such technically advanced aircraft that they were able to be used efficiently until the end of the war.

One of the most extraordinary examples of these aircraft was the 'schnellbomber' Junkers 88 with its ability to adapt equally well to the most varying roles.

Born as a bomber, it became the most important German night fighter, and thus one of the main pillars in the defense of the Reich.

The history of the Ju 88 fighter begins with one of the bomber prototypes, the V7, tested by the company as a 'zerstörer' shortly before the beginning of the war. Not very convinced, the RLM accepted the proposal of the Ju 88c version, a heavy fighter with a solid nose and a fixed armament of two 20 mm MG/FF guns and two 7.9 mm MG17s.

Production got under way in 1940 with the C-2 type, fitted with 1200-HP Jumo 211B-1 engines, and carried on very slowly until the beginning of 1942 with the succeeding C-4 and C-5 models. The latter was fitted with BMW 801 radial engines (already installed previously in the only C-3) which, thanks to their greater power, improved performance even more. In the spring of 1940,

the Luftwaffe used these aircraft to initiate night intruder missions against British bombers and Bomber Command bases, while suitable tactics were being worked out for defensive use. However, priority for defensive night fighters was given to the Bf 110. In fact, at that time it was felt preferable to take away as few Ju88 fighters as possible from the bomber production lines.

Only at the outset of 1942 did more substantial supplies of Ju88 fighters begin with the C-6; it was practically identical to the previous models, fitted with 1400-HP Jumo 211Js and widely used, going to supplement the previous types already in use in the Mediterranean and over the French Atlantic coast too. At the end of '42 the C-6s appeared in a specialized night fighter version fitted with the new Telefunken radar, FuG 202 Lichtenstein BC and FuG 212 Lichtenstein C-1. These aircraft, together with the Ju 88R-1s (which differed only by their 1600-HP BMW 801 engines), the much more numerous Bf 110s and a small number of Do 217s, were fitted into a well-organized defense system, thanks to which the losses inflicted on the British night bomber formations rose to worrisome levels.

The British use of that simple and effective

A Ju88G-1 tested by R.A.F.

countermeasure known as 'window' from July 1943 onward blinded the German radars operating on a frequency of 90 megacycles, and it was only some months later that the Ju 88s and other night fighters received the new FuG 220 Lichtenstein SN-2s which operated on 490 mega-cycles. Meanwhile, the alternative 'wild boar' technique had been developed, based on a strong illumination of the area to be defended in order to permit visual contact between the German fighters and the attacking bombers.

The Ju 88 with SN-2, was designated C-6c and was armed with 3x20 mm and 3x7.9 mm guns in the nose; it proved itself to be a very effective night fighter, so much so that many officials fett

A post war photo of a Ju88R exhibited in England.

no pressure to re-equip units with other more modern types, more specialized and perhaps even better, such as the He 219. In fact, they preferred development of other versions of the excellent and well-tested Ju 88 airframe.

The Ju 88C-6c was largely responsible for a real massacre of British bombers during the early months of 1944, and as a consequence of this, the RAF had to interrupt their night air offensive and wait for effective countermeasures.

Meanwhile, of course, the development of further versions continued: the Ju 88 C-7 for day use as a 'zerstörer', while the new night-fighter models incorporated numerous modifications (the most obvious being the adoption of a larger tail assembly, similar to that of the Ju 188), enough to justify the assignment of the new designation Ju 88G. This version used the 1700-HP BMW 801D radial engine, which was fitted in the G-1 and G-4 models, while the first series of the

G-6 got the BMW 801G. The last G-6s and the G-7s (the last model of this version produced) received the Jumo 213A and 213E liquid-cooled engines of around 1700 HP.

The Ju 88G started its career in 1944, so right from the beginning it used Lichtenstein SN-2, as well as getting other sophisticated equipment for the 'electronic war' such as the FuG 227 Flensburg apparatus for surveying the emission of the British bombers' tail warning radar.

In the meantime the British carried on their relentless efforts at perfecting new devices for jamming the German radar, and this fact caused, in the Ju 88Gs during the last months of the war, a continual process of adaptation of the electronic equipment to more and more sophisticated requirements: FuG 228 Lichtenstein SN-3 and FuG 218 Neptun variable frequency radar, FuG 240 Berlin centimetric radar, and devices for surveying enemy radar emission — the already recalled Flensburg, and the FuG 340 Naxos.

Armament in the Ju88G was generally composed of four MG151 guns in a ventral tray, two more in a Schräge Musik arrangement, and one 13 mm MG131 on a flexible mount for rear defense.

At a certain point the electronic equipment became so complex as to actually require two operators, thus bringing the crew number to four. After the great victories in the spring of 1944, the efficiency of the German night fighters decreased relentlessly, due to the effectiveness of the British countermeasures (facilitated by the fortuitous capture of intact examples of the Ju 88C, R and G), the worsening of the military situation in general, and, finally, the problems in training new crews to replace those lost in battle. Despite this the Ju 88 was, right to the end, an adversary to be feared by the British bombers, and the performance of the last versions reached unusual heights for aircraft of those dimensions and weight: its maximum speed had in fact reached around 640 km/h (400 mph). All the fighter versions of the Ju 88 came to the considerable total of about 4000 examples built, with just under 3000 in 1944 alone!

The successor for the Junkers 88 was to have been the Ju 388, the last extrapolation of this successful twin of late '30s vintage. The new aircraft was optimized for use at high altitude and all its expected versions — bomber, reconnaissance and all-weather fighter — had pressurized crew compartments.

The all-weather fighter was designated Ju 388J and the first prototype was the Ju 388 V2, flown at the beginning of '44. Another two prototypes followed, the V4 and V5, and this brought to an end the productive career of the promising fighter. In fact, production of the first J-1 series

versions, already under way, was halted by the vast program of potentiation and rationalization of the German fighter forces, known as the Emergency Fighter Program, in November 1944. The Ju 388 had been cancelled and the directions were respected, but this was not so in other cases, such as for the He 219, where orders were tacitly ignored.

The engines of the prototype of the Ju 388J were the turbo-supercharged BMW 801-TJs rated at 1400 HP at 12,300 m (40,300 ft), and at the same altitude maximum speed was 580 km/h (360 mph). The crew was four; there were 2x30 mm + 2x20 mm cannon in a ventral fairing and 2x20 mm Schräge Musik guns. The defensive armament was to include 2x13 mm MG 131s in a remotely-controlled tail turret, experimented on the first prototype but which was to be available in series only in the Spring of '45.

The radar was FuG 218 Neptun whose antennas were contained almost completely in a streamlined wooden nose; the combat avionics were completed by Naxos.

The later versions of the Ju 388 were intended to use liquid-cooled engines of the Jumo 323 and the Daimler Benz DB series.

The Ju 388 V-2 nose with FuG220 radar array.

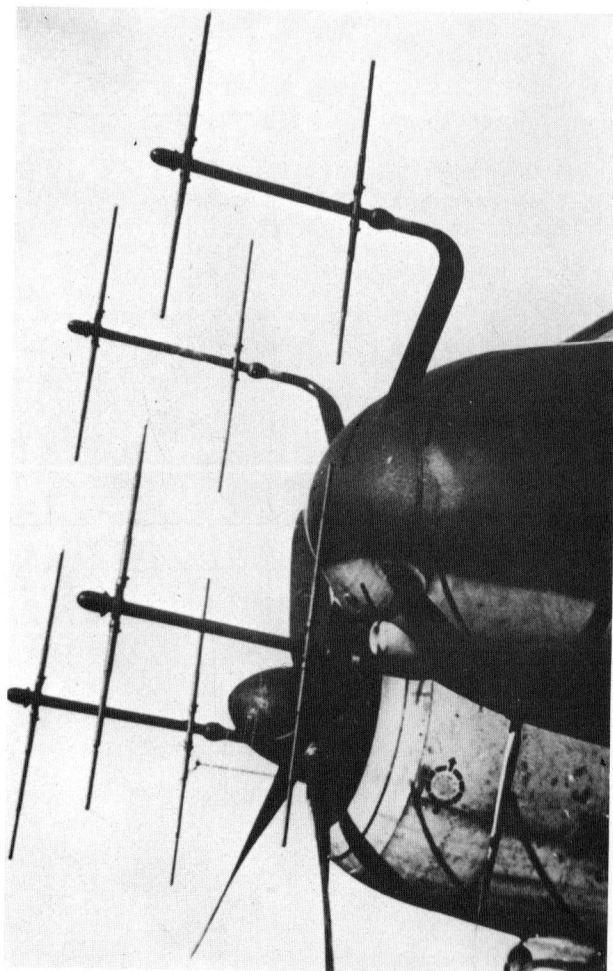

Junkers Ju 88 C-6c

Three-seat night fighter

Power plants: two Junkers Jumo 211 J-1 or J-2, 12-cylinder liquid-cooled engines, each rated at 1340 HP for take-off, driving three-blade controllable-pitch VDM airscrew.

Dimensions: span 20 m; length (excluding radar array) 14.36 m; height 5.07 m; wing area 53.50 m².

Weights: empty equipped 8060 kg; normal loaded 12350 kg.

Performance: maximum speed 494 km/h at 5300 m; cruising speed 423 km/h at 6000 m; climb rate 9 m/sec at sea level; ceiling 9900 m; normal range 1040 km.

Armament: three 20 mm MG/FF cannon with 360 rpg; three 7.9 mm MG17 machine-guns with 2800 rounds; two 20 mm MG 151 cannon with 400 rpg in a 'schräge musik' installation; one 13 mm MG 131 machine-gun on a flexible mount for rear defense.

JU 88C-6C

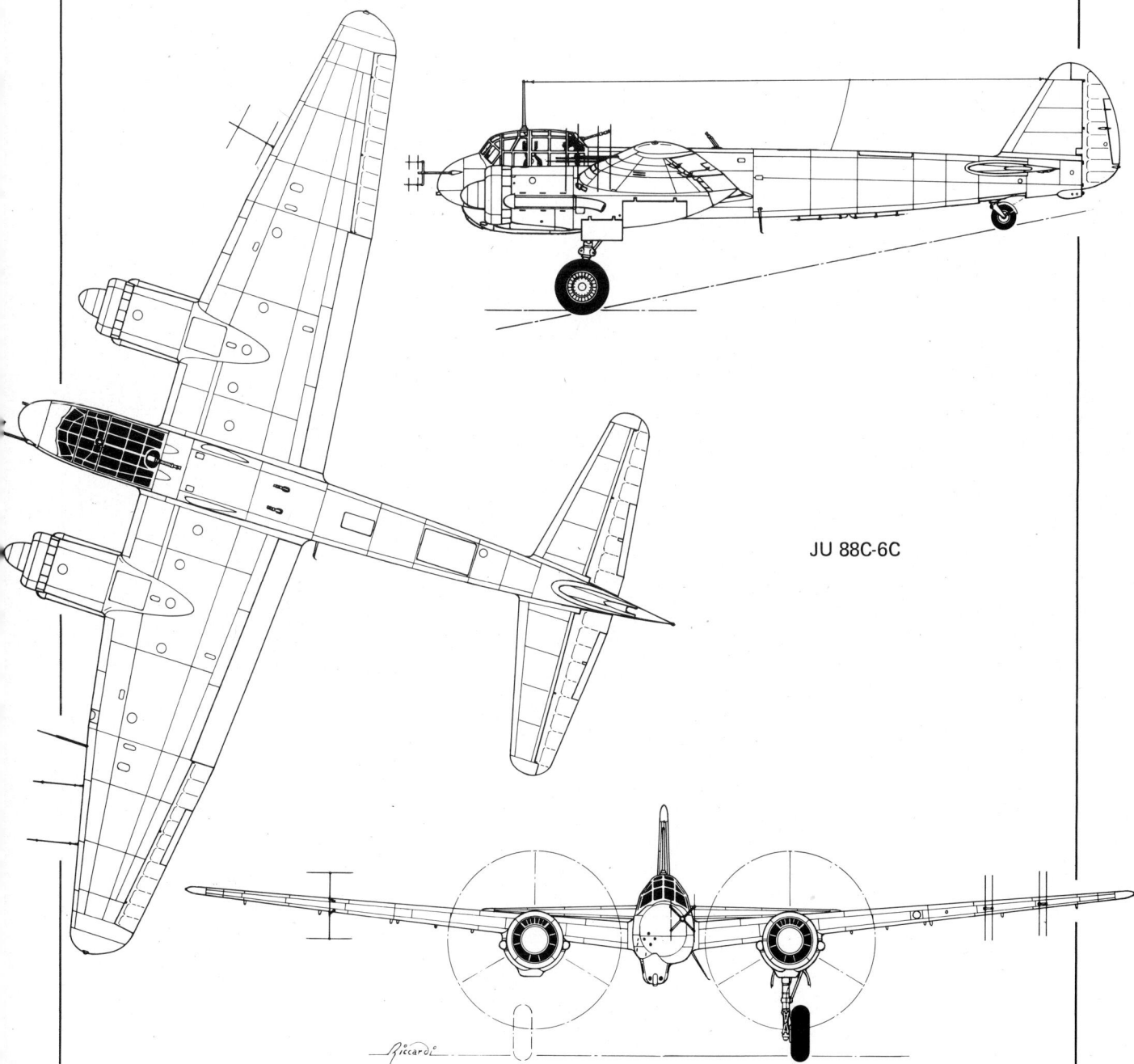

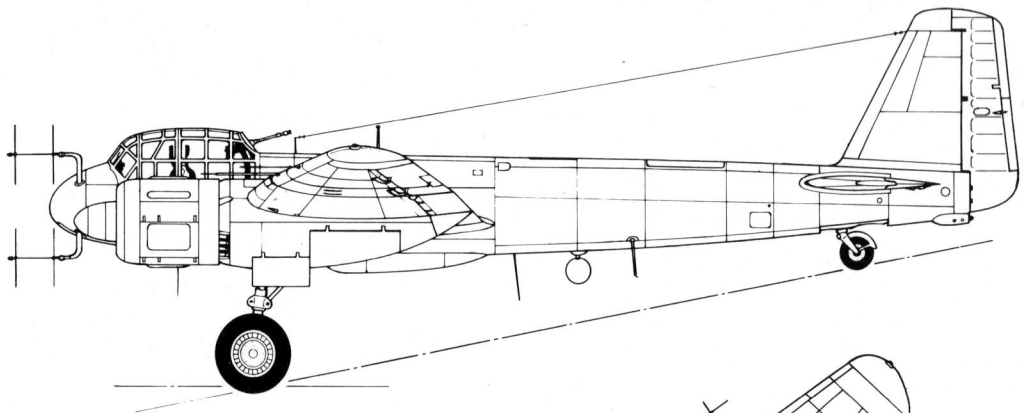

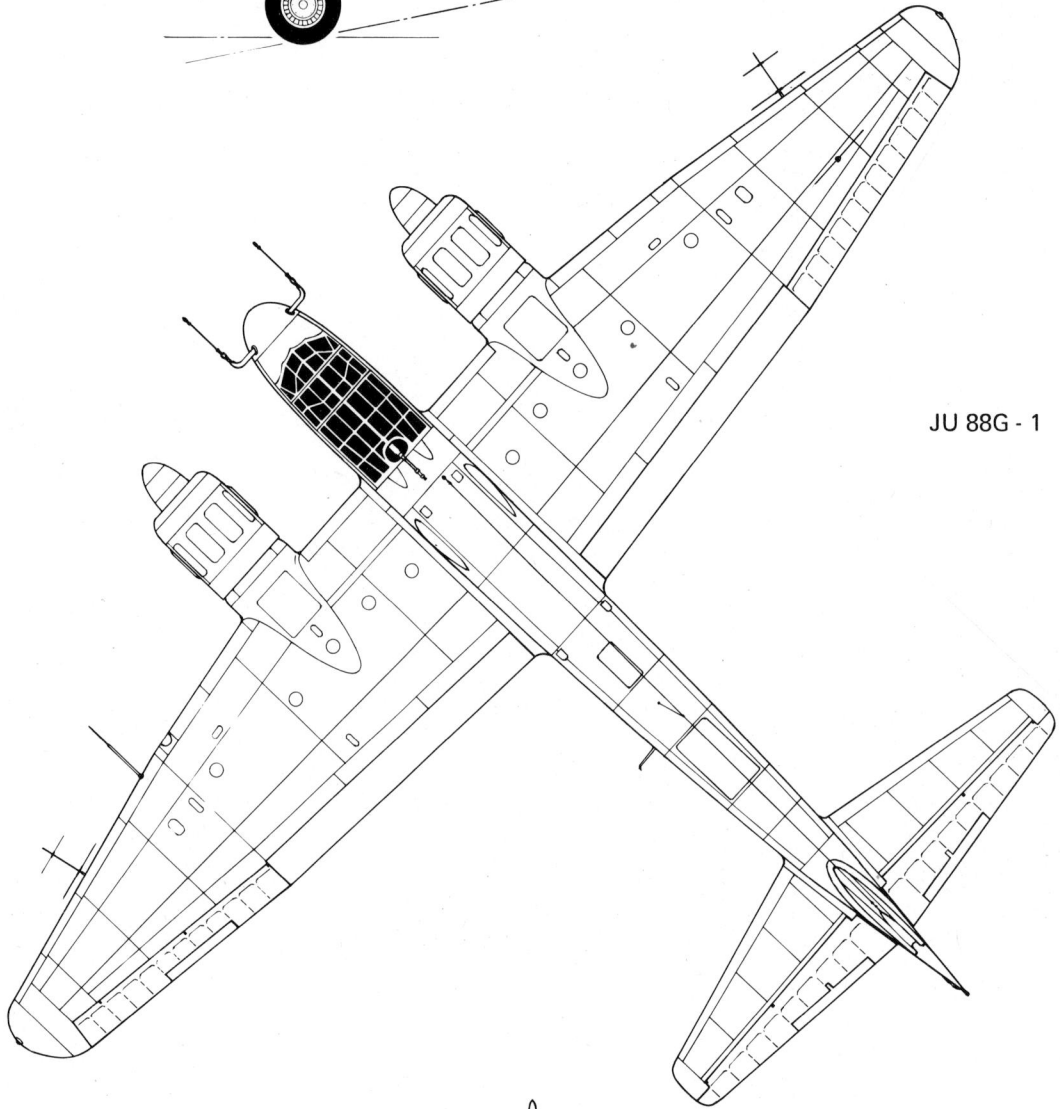

JU 88G - 1

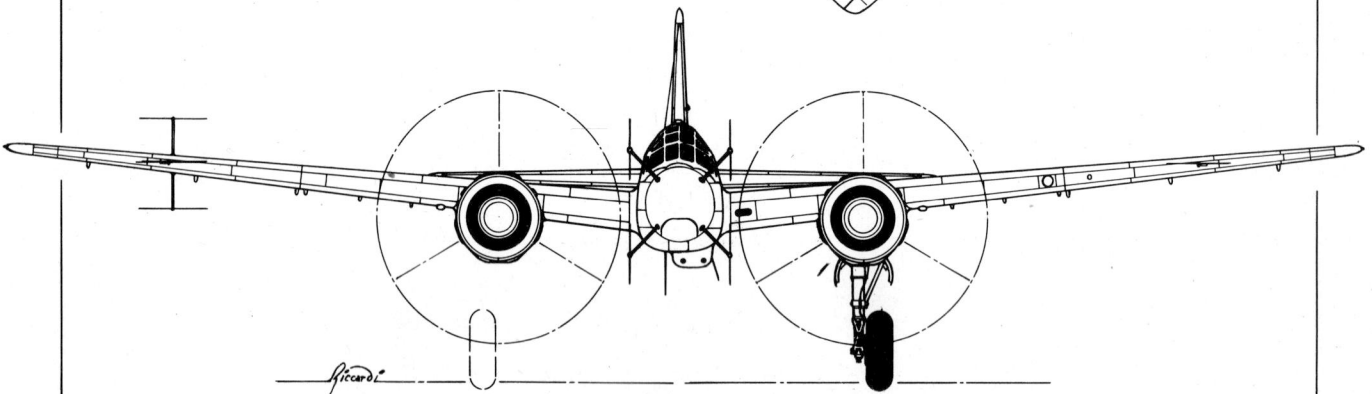